T0261115

The Future of Creative Work

The Future of Creative Work

Creativity and Digital Disruption

Edited by

Greg Hearn

*Professor, Creative Industries Faculty,
Queensland University of Technology, Australia*

Edward **Elgar**
PUBLISHING

Cheltenham, UK • Northampton, MA, USA

Published by
Edward Elgar Publishing Limited
The Lypiatts
15 Lansdown Road
Cheltenham
Glos GL50 2JA
UK

Edward Elgar Publishing, Inc.
William Pratt House
9 Dewey Court
Northampton
Massachusetts 01060
USA

A catalogue record for this book
is available from the British Library

Library of Congress Control Number: 2020942779

This book is available electronically in the **Elgar**online
Business subject collection
http://dx.doi.org/10.4337/9781839101106

ISBN 978 1 83910 109 0 (cased)
ISBN 978 1 83910 110 6 (eBook)

Printed by CPI Group (UK) Ltd, Croydon CR0 4YY

Contents

Figures

Tables

Contributors

Müge Belek Fialho Teixeira is a creative maker and trans-disciplinary designer, and a team member of Queensland University of Technology Design Robotics, specialised in advanced manufacturing, digital fabrication and parametric design. She has worked with Zaha Hadid Architects, taught in several institutions including AA Visiting Schools, published articles, given interviews, and presented at many international conferences and exhibitions. Müge has also been awarded in multiple competitions and events on the future of architecture and the use of novel digital technologies.

Ana Bilandzic is a PhD candidate with the Urban Informatics Research Group at the Queensland University of Technology. Her research is on social and spatial precursors to innovation in casual creative environments. Ana is unpacking those precursors in innovation hubs that have a focus on social innovation in the peri-urban area of Brisbane, Australia. Her research is motivated by the increasing number of innovation spaces around the world and unmet user needs for social and thematic diversity in such spaces.

Chris Bilton is Reader in Creative Industries at the Centre for Cultural & Media Policy Studies, University of Warwick. He is author of several books and articles on management in the creative industries, creative strategy and 'uncreativity', including *The Disappearing Product* (2017), *Management and Creativity* (2007) and *Creative Strategy* (with Professor Steve Cummings) (2010). Before moving into academia, Chris worked in community arts and fringe theatre in London. He is currently working on a co-authored book on multiple creativities.

Cliff Bowman, Professor of Strategic Management, Cranfield School of Management, has spent his academic career trying to help managers develop their strategic leadership capabilities. In this pursuit, he has developed techniques that managers seem to find useful, which are underpinned by sound theory about the processes of value creation, value capture and strategic change. Recent books include: *Embracing Complexity* (with J. Boulton and P. Allen) (2015, Oxford UP) and *What's Your Competitive Advantage* (with P. Raspin) (2018, FT Pearson).

Ruth Bridgstock is Professor and Director (Curriculum and Teaching Transformation) in Learning Futures at Griffith University. Ruth is passionate about fostering future capability in learners, teachers and educational institutions. She leads research and scholarship into the changing world of work and social challenges we all face, capability needs, and approaches to learning in the digital age. Ruth is Principal Fellow of the Higher Education Academy (now Advance HE UK), and Australian National Senior Teaching Fellow. Her blog is at futurecapable.com.

Glenda Amayo Caldwell is a senior lecturer in architecture at the Queensland University of Technology. She is a chief investigator in the IMCRC Design Robotics for Mass Customisation Manufacturing project in collaboration with UAP and RMIT University. Glenda embraces trans-disciplinary approaches from architecture and human–computer interaction and her research places design at the forefront of robotic research for design-led and mass customisation manufacturing. She is the author of numerous publications about media architecture, urban informatics and design robotics.

Natalie Collie lectures in media studies and professional writing in the School of Communication and Arts, University of Queensland. Her research includes a focus on the future, science fiction, the body, urban space and digital cultures.

Stuart Cunningham AM is Distinguished Professor of Media and Communications, Queensland University of Technology. He is a fellow of the UK Academy of Social Sciences and the Australian Academy of the Humanities, and a Member of the Order of Australia. His most recent books are *Social Media Entertainment: The New Intersection of Hollywood and Silicon Valley* (with David Craig) (2019, New York University Press) and *Media Economics* (with Terry Flew and Adam Swift) (2015, Palgrave).

Greg Hearn is Professor and Director of Commercial R&D in the Creative Industries Faculty at the Queensland University of Technology. He researches and writes on the role of the creative industries in the knowledge economy, the creative workforce across all industries, and technology/work futures in general. Greg is a research leader in the Advanced Robotics for Manufacturing Hub at the Queensland University of Technology, with a focus on the role of design, and the impact of robotics on work.

Rui Oliveira Lopes is Assistant Professor of Art and Design History and Programme Leader for Design and Creative Industries at the Faculty of Arts and Social Sciences, Universiti Brunei Darussalam. His research interests include global art history, artistic and cultural exchange between Europe and Asia in the Early Modern Period, cultural and creative management, and the role of museums and artistic practice in cultural negotiations. Rui has curated exhibitions and collaborated with several museums in Europe and Asia.

Marion McCutcheon is a research associate with the Digital Media Research Centre at the Queensland University of Technology and Honorary Research Fellow at the University of Wollongong's C3P Research Centre for Creative Critical Practice. As a communications economist, she has worked within the Australian Government in telecommunications and broadcasting policy advisory and research roles. Marion's current research includes examining the role of the creative industries in economic systems, and how society benefits from the making and use of cultural artefacts.

Onur Mengi is an inter-disciplinary academic teaching industrial design and an inter-disciplinary designer working on projects ranging from product design to urban design. He is Assistant Professor in Industrial Design and Vice Dean of Faculty of Fine Arts and Design at Izmir University of Economics in Turkey, and a research fellow in the Creative Industries Faculty at the Queensland University of Technology in Australia. Onur's research concentrates on urban design, urban planning, design strategies, design management and place-based development processes.

Jason Potts is Professor of Economics in the School of Economics, Finance and Marketing at RMIT University, and Director of the Blockchain Innovation Hub. Jason is a Fellow of the Academy of Social Sciences of Australia and one of Australia's leading economists on economic growth, innovation and institutions, and on the economics of cities, culture and creative industries. He is editor of the *Journal of Institutional Economics*. His latest books are *Innovation Commons* (2019, OUP) and *Understanding the Blockchain Economy* (2019, Elgar).

Andreas Pyka is Chair for Innovation Economics at the University of Hohenheim and researches neo-Schumpeterian and evolutionary economics, with an emphasis on numerical techniques of analysing economic dynamics. His research focuses on knowledge-driven developments and transformation of economic systems towards sustainability.

Ellie Rennie is a principal research fellow in the Digital Ethnography Research Centre at RMIT University and a member of the Blockchain Innovation Hub. She investigates digital inclusion and automation, with a focus on public policy. From 2020, Ellie will be an Australian Research Council-funded Future Fellow. Her books include *Using Media for Social Innovation* (2017, Intellect) and *Internet on the Outstation: The Digital Divide and Remote Aboriginal Communities* (2016, INC).

Jonathan Roberts is Professor in Robotics at Queensland University of Technology and researches field robotics, design robotics and medical robotics. Previously, Jonathan was Research Director of CSIRO's Autonomous Systems Laboratory and was a co-inventor of the UAV Challenge, an international flying robot competition. He is also a past president of the Australian Robotics & Automation Association, was a member of the Founding Editorial Board of the *IEEE Journal of Robotics and Automation Letters*, and currently serves as a senior editor.

Jose Hilario Pereira Rodrigues is a research fellow at the Queensland University of Technology and the Queensland Academy of Sport. His work focuses on ecological skill acquisition, expertise and performance, and he divides his time between researching dance and sport. Jose has widespread international experience, delivering high-quality activities and programs in Portugal, England, Germany, Singapore and Australia. Among his other interests are management, coaching, psychology and educational business models that can facilitate adaptation, creativity and innovative capacity of individuals.

Mark David Ryan is an associate professor in screen and media industries and a chief investigator for the Digital Media Research Centre at the Queensland University of Technology. He is an expert in screen industries research, Australian genre cinema, and digital media. He was President of the Screen Studies Association of Australia and Aotearoa/New Zealand (SSAAAZ) between 2015 and 2018 and an executive member of the Australian Screen Producers Education and Research Association (ASPERA) in 2015/2016.

Pier Paolo Saviotti is a visiting professor in Innovation Studies at the Copernicus Institute of Sustainability, Utrecht University, and at the St Anna School of Advanced Studies, Pisa. His research interests are in evolutionary economics, covering the economics of innovation, of development and of knowledge.

Aljosha Karim Schapals is a lecturer in journalism and political communication in the School of Communication at the Queensland University of Technology in Brisbane, Australia, as well as a research associate in the Digital Media Research Centre, where he is working on the Australian Research Council Discovery Project 'Journalism beyond the crisis'. He is the lead editor of *Digitizing Democracy*, a major edited collection in the field of political communication, published by Routledge in 2019.

Cori Stewart is a strategic leader who co-innovates with industry, government, research institutions and the community. She is an associate professor in creative industries at Queensland University of Technology with research expertise in public policy. Cori is CEO of the Advanced Robotics for Manufacturing Hub (ARM Hub), which specialises in advanced robotics and design-led manufacturing and is delivering Industry 4.0 projects with various companies and research institutions. The ARM Hub assists micro, small and medium enterprises on their Industry 4.0 digital transformation journeys.

Juani Swart is a professor in human capital management at the University of Bath and seeks to understand the nature of contemporary work contexts. She is particularly interested in cross-boundary working (including gig-working and artificial intelligence) and its impact on commitment and knowledge sharing. Juani's research is focused on human capital as a strategic resource, innovation, ambidexterity, and employee attitudes and behaviours. She has published widely in journals such as *Human Resource Management, Human Resource Management Journal, Organisation Studies* and *Journal of Management Studies*.

Beverly Yuen Thompson is an associate professor of sociology at Siena College, in Albany, New York. She has written on gender and subcultures, including her monograph *Covered in Ink: Tattoos, Women, and the Politics of the Body* (2015, NYU Press), women digital nomads and social movements. She earned her PhD from the New School for Social Research in New York.

Russell Tytler is Alfred Deakin Professor of Science Education at Deakin University. He has researched and written extensively on student learning and reasoning in science. Russell's interest in the role of representation in reasoning and learning in science extends to pedagogy and teacher and school change. He researches and writes on student engagement with science and mathematics, school–community partnerships and STEM policy.

Ben Vermeulen is a post-doctoral researcher at the Eindhoven University of Technology and the University of Hohenheim, Stuttgart, and prior to this, he worked for a decade as a software engineer in industry. His research interests are in evolutionary economics, structural change, the geography of innovation, sustainability transitions, innovation policy and agent-based modelling.

Peta White is a senior lecturer in science and environmental education at Deakin University. Peta has worked in classrooms, as a curriculum consultant and manager, and as a teacher educator in several jurisdictions across Canada and Australia. She gained her PhD in Saskatchewan, Canada, where she focused on learning to live sustainably, which became a platform from which to educate future teachers. Peta's current research interests include science and biology education; sustainability, climate change, and environmental education; and collaborative/activist research.

Caroline Wilson-Barnao is a lecturer in the School of Communication and Arts at the University of Queensland. She has many years' experience in public relations and marketing, especially for non-profit organisations, and currently teaches in theoretical and practical subjects. Caroline's research takes a critical focus on the use of digital media in cultural institutions and public space.

1. The future of creative work: creativity and digital disruption

Greg Hearn[1]

INTRODUCTION

What does the future hold for creative work? What do we actually mean by the term *creative work*? Do we only include work in art, media and design organisations? What about artists working in community services, videographers in health promotion, experience designers in banks, art directors selling toothpaste? Is publicly funded creative work different from commercially funded creative work? Should we in fact include any job that contains creative problem solving? Or do we need to think more fundamentally about the difference between scientific creativity and artistic creativity? And then there is the question of technology. Are creative jobs really safe because robots cannot be creative? How are algorithms automating tasks that were previously thought to be creative, for example, in journalism? How are digital platforms changing the creative value chain and, in doing so, changing creative employment, how value is created and who captures it? Such questions underlie the two main contentions of this book. First, there is a major disjunction between the future world of creative work as imagined by aspiring undergraduates, promoted by their teachers and critiqued by researchers, and the world of creative work as it is actually enacted. Secondly, enmeshed with society and culture, technological *evolution* is accelerating, perhaps even toward a radical *step-change*. Therefore, the institutions, organisational structures, regulations, beliefs, motivations, identities and habitus that govern and enable creative work are changing, perhaps even radically.

The complexity of *defining* the creative work landscape occurs because culture is no longer only invested in stand-alone cultural artefacts, but also interpenetrates every product, service or experience we have. Therefore, it can be argued that creative work exists not just in organisations devoted to producing stand-alone cultural artefacts, but also in virtually every organisation, public or private, in every sector, producing any good or service. This is the creative economy at work. In this book, the term *creative* as applied to skills or work has a very specific definition. In Chapter 2 of this book, Hearn and McCutcheon (2020) propose that work or skills are deemed creative if they comprise "replicative or novel aesthetic and/or expressive knowledge, either separately or in synthesis with other forms of knowledge" (p. 17). The *creative economy* is conceptualised in this book as the creation capture and consumption of intangible value through the application of creative, techno-logical and innovation know-how. A key element in this conceptualisation is *intangible value*. Haskell and Westlake (2018) make the point that, from the beginning of the twenty-first century, developed economies began to invest more in intangible long-term productive assets (e.g., knowledge, research, design, branding, software, market relationships and consumer metrics) than in traditional tangible assets (e.g., buildings, facilities or machinery). They point out that intangible investments are made by companies across many kinds of value chains. They are also made by governments, for example, in research and development (R&D). This claim is not mere rhetoric. The meas-urement of intangible assets is a detailed financial field, not least because such investments need to be managed as part of prudent financial management and external stock valuations, and as part of managing national economies (Haskell and Westlake, 2018). In this book, the term *intangible value* encompasses intangible assets, but also includes more ephemeral intangibles such as digital content, advertising copy or performance improvisations that are unlikely to be re-used. With the rise of digital platforms, understanding intangible value in the creative economy is even more important, whether you want to understand *or* critique the creative economy.

While most intangible creative assets are held by private or publicly listed companies, perhaps the more complex question is who benefits from intangi-ble value in creative value chains. Increased creative employment is a benefit that could result as much as increased profit. This question goes to the heart of issues of creative work conditions in the future, as digital evolution proceeds to disrupt all stages of creative value production and consumption. In future, cre-ative intangible value could be increasingly codified into algorithms, systems and assets (cf. Bowman and Swart in Chapter 12) owned by global monop-olistic companies. But do consumers have power, and to what extent would such a standardised creative landscape be acceptable? Moreover, in general, across all products and services, diversity and customisation is increasing. In this scenario, the richness of hard-to-replicate creative capabilities embodied in workers (Bowman and Swart, 2007) could become a competitive advantage. As any technological evolutions play out, at least in the short term, there are winners and losers. Similarly, there will be winners and losers in the creative work ecosystem. But who will they be?

Intangible capital growth and digitisation are challenging every creative job. The digital drivers of future technological evolution that affect the future of creative *work* include not only the purely digital (e.g., software, platforms, applications, machine learning, artificial intelligence (AI), chatbots), but also digitally enabled mechatronic agents (e.g., 3D (three-dimensional) printing, drones, autonomous camera systems, robots). In the realm of creative work, software and mechatronic agents are learning to perform increasingly complex and, arguably, creative tasks through AI, machine learning and biomimetics. It is easy to overestimate how effective these systems are in terms of creative work, but certainly they are becoming prevalent. All steps in creative economy supply chains are affected by these forms of digitisation, including:

- the design process of products and services (e.g., augmented reality, robotic prototyping);
- the formulation, production or manufacturing and packaging of products and services (e.g., drone footage, robo-journalism);
- attracting consumers to these products (e.g., recommendation algorithms);
- managing the consumption of those products and services (e.g., apps for health care, service chatbots).

The scope of changes to creative work is large and the process of change complex because technology per se is not the only driver of change. In concert with this technological evolution are significant patterns of social, institutional and cultural evolution in which the technology is tested, leading to either radical or incremental adaptations or resistance. This means the kinds of organisations in which creative work will be found in the future are also evolving. Different sectors of the creative economy (e.g., cultural production, creative services) have a different constellation of organisations of different sizes. Which creative sector you work in will determine many of the characteristics of the work involved. Organisational changes will affect the extent to which work is formal, or informal and portfolio based. Digitalisation involves trends to casualisation and contract-based work away from permanent positions. Equally, some creative organisations adopt a successful strategy of using permanent positions to attract and retain high-value creative workers. Autonomous systems can mean that fewer people are needed in routine pro-duction; however, more work is needed to design, produce and manage the autonomous system.

Given all of the above, we also need to think about the kinds of skills or capabilities required for the future of creative workers. If creative work involves *replicative or novel aesthetic and/or expressive knowledge, either separately or in synthesis with other forms of knowledge,* which of these elements will be most important in any given work context? Will all creative workers need digital, visual and communicative literacy? Will creative work be predominately undertaken in creative teams of discipline specialists or do individuals need to be multi-skilled? What education is needed to enable multi-disciplinary capacities and/or team management and an understanding of the creative process for teams versus individuals? Required skill sets could also depend on what part of the creative value chain is involved. Is the work more in the design phase, or in production, or is it customer facing? Is it in the art, media or design sectors, or in healthcare, education or tourism? Is the work done in a small specialist agency, a large government bureaucracy or by engaging a sole-trader? In addition to creative literacies, given the social, institutional and cultural adaptations likely in the creative work of the future, how important is an understanding of these forces? How important is an understanding of the social and technological regulatory environment, and business or legal skills? Skills in ethics and worker rights advocacy and organisation are needed to negotiate the changing institutions that govern creative work. Over and above specific disciplinary skill sets, there will be a need for career management skills to build professional networks, negotiate pathways into and around the creative labour market, and manage career transitions.

The considerations and questions raised so far give rise to the organisation of this book in four themes:

1. The evolution of creative work.
2. Digital disruption and creative work.
3. Changing contexts of creative work.
4. Educating for the future of creative work.

THE EVOLUTION OF CREATIVE WORK

In Chapter 2, Hearn and McCutcheon define the creative economy with particular attention to the growth in intangible capital and its interdependence with human capital. Creative work features in several of these intangible assets, namely, software and digital content, entertainment and arts original works, design of products and services, and advertising and marketing collateral. With reference to standard national occupational codes, the distinction between cultural production and creative services occupations is defined, as well as the distinction between creative work in creative firms and creative work in other sectors of the economy. This model of the creative economy is exemplified empirically using employment data from the Australian Census. It is likely that the future of creative work is different depending on the type of occupation and the sector of deployment. Moreover, the creative economy creates opportunities for some, but also disrupts power relations in the world of work, potentially leading to new axes of disadvantage within the creative workforce. Compared to the average Australian worker, creative workers are in general better off in terms of job growth rates and salaries. The exceptions are in publishing and visual arts. And within the creative workforce, there are issues of inequity, particularly in relation to part-time work, women and non-urban locations. The increasing returns dynamic of the creative economy means that urban locations, with their intensity of capital, higher concentration of creative workers and spill-over application in other sectors, will continue to be predominant in the future of creative work. However, the creative economy extends beyond major cities.

To pursue this issue, in Chapter 3, Hearn, Cunningham, McCutcheon and Ryan report on exploratory quantitative and qualitative investigations of the creative economy outside the capital cities of Australia. This work focuses specifically on the relationship between creative employment and Gross Regional Productivity (GRP). This chapter utilises a national data set of all 478 local government areas (LGAs) in Australia, differentiating between cities, inner regional, regional, outer regional, remote and very remote areas. In addition, this chapter presents qualitative accounts of a remote distressed economy with a population of 14 000, a stable economy with a population of 150 000 and a strongly growing economy with a population of 350 000. Hearn et al. argue that creative economy activity can be found in LGAs from second-tier cities to remote communities, and that creative workers make valuable local economic contributions both when economies are distressed and when they are booming. Not only are there more creative jobs in local economies with higher GRP, but the creative intensity (i.e., creative occupations' share of total employment) also increases with size. However, the creative economy operates with different dynamics in regions with different size populations, GRP and distances from major cities. For example, creative intensity in software and digital content and in advertising and marketing correlates with GRP across all types of regions. Film, TV and radio make a surprising economic contribution in remote communities based on film location and film tourism, and the correlation between the creative intensity of music and performing arts and GRP in outer regional areas is higher than it is in cities.

In Chapter 4, Vermeulen, Pyka and Saviotti further the discussion of the evolution of creative work by theorising the relationship between creative employment and the rise of robotics and artificial intelligence (R&AI). They point out that, to date, robots and software have been designed to displace workers doing routinised, computationally intensive or dangerous work. However, this is only one mechanism of relevance to the future of creative work. Vermeulen et al. develop a taxonomy of the different types of industrial engagement with robotics based on whether R&AI are made (i.e., through R&D, manufacture and servicing), applied (in both creative and non-creative economy sectors), or supported (e.g., through training, education, consulting). They propose ways in which employment, tasks performed and income differ in each type of engagement. Vermeulen et al. argue that tasks requiring creative skills (together with tasks requiring social skills and dexterity) may become more prominent in the future workforce. In addition, they suggest that application of R&AI in creative sectors may create whole new industry segments, in the same way that the games industry developed from computing applications. This chapter then develops three scenarios on the future of employment and R&AI: the *end of work*, *employment rebounds* and *lower level total employment*. This chapter also discusses the role of institutions (e.g., labour unions, social security, education), and structural shifts in countries' competitiveness in each scenario. The authors suggest that the development of employment is affected by the scale and scope of applicability of R&AI, the educability of people and the innovativeness of economies. The chapter ends with a critical note on the polarisation of, and inequality in, labour markets within and across countries.

DIGITAL DISRUPTION AND CREATIVE WORK

In Chapter 5, Rennie and Potts discuss the negative impact of digital disruption, particularly streaming services, on cultural producers such as musicians, artists and film makers, in terms of time imposts and payment for their work. Australian artists overwhelmingly identify intrinsic factors as drivers of career advancement. But they experience barriers such as time and monetary costs in accessing specialised business, legal and technology knowledge. In addition, there are challenges beyond their control, particularly industry changes in audience, buyer behaviour and payment systems. In response, Rennie and Potts propose a radically new *industry utility* model, based on blockchain technology, as a new economic infrastructure for cultural producers. Blockchain allows human or non-human agents to move value through the internet. For creatives, this value could be money, equity tokens or creative works themselves. This distributed ledger ensures that value cannot be copied or spent twice and therefore provides, in theory, a trustworthy means of exchange of ownership and/or payment. Although there are uncertainties in the development of blockchain technologies, such a system, if realised, could make the creative economy more transparent and efficient, encourage peer-to-peer markets, and remove intermediaries between creative workers and their fans. If established, the system could provide consumer benefits through open competitive access to the system, enhance the total productivity of the creative industries, and allow cultural producers to spend more time on creative content production.

Schapals discusses the impact of so-called *robo-journalism* on the future of journalism in Chapter 6. Since the late 2000s, algorithms have been used to create narratives from large volumes of data, for example, for sports and financial results. In some studies, readers have been unable to differentiate between machine and human authors, though evaluations of specific criteria show that humans produce narratives that are perceived as coherent, readable and pleasant to read, while algorithms are perceived as more credible and objective. Algorithms can reduce the need for fact checking and greatly reduce production costs. However, they require clean, structured and reliable data to be effective. Also, algorithms cannot yet infer causality nor interrogate data for meaning, so they cannot produce the more complex investigative articles that shape public opinion. Nevertheless, the rise of robo-journalism raises critical questions for the future of journalists' work. Will it make journalists expendable or free up their time to engage in more creative and critical writing? Will journalistic accountability, ethics and transparency be a source of competitive advantage for journalists in the race against the machine? In an era in which global numbers of print journalists are contracting, this chapter raises key questions about the precarity of journalism, and suggests opportunities that could provide a countertrend in terms of high-value investigative capabilities.

Stewart, Caldwell, Belek Fialho Teixeira and Roberts argue in Chapter 7 that digital technologies create unprecedented possibilities for architects to create new forms, and to build places and spaces more suited to the Anthropocene epoch. R&AI can now be found in every stage of the architectural design process and, concomitantly, in construction of the built environment per se. For example, R&AI enables more explorative design and rapid prototyping, mass customisation, accelerated modular construction and up-front estimation of the environmental impact of design choices. At the same time, digital technologies are poised to have a disintermediating effect on the profession of architecture through Uber-like services that allow customers and builders to pick and choose different design elements. This potentially changes the business model and traditional industry structure of specialist architecture firms, which to date have proven very resilient. Nevertheless, Stewart et al. argue that experts in the "future of work" indicate that architecture is one of the jobs least likely to be automated in the future due to hybridity in the complex skill set required. Architecture requires digital design, creativity, engineering, and cultural knowledge, making it less susceptible to automation than most occupations.

Museums are longstanding and often publicly funded cultural and educational institutions. Lopes argues in Chapter 8 that the museum has undergone significant changes in recent decades and an accelerating path of digitisation is expected. Focusing on art museums, this chapter discusses the interactions between museum curation and digital media, and the effects on the future of creative work in these fields. Lopes argues that many museums have in fact evolved as kinds of incubators using technology to enhance the interpretation, presentation and curation of collections. The rapid growth of digital media is having a profound effect on the work of curators and other museum professionals. Collaboration with other creative workers, from designers to social media strategists, facilitates new venues, new audiences, and innovative visitor experiences in and beyond the confined physical spaces of the museum. This chapter proposes an expanded notion of museum curation that goes beyond the traditional role of the curator and the physical space of the museum institution.

CHANGING CONTEXTS OF CREATIVE WORK

In Chapter 9, Bilandzic, Mengi and Hearn discuss the factors leading to the proliferation of casual creative environments (CCEs) such as co-working spaces, innovation hubs, incubators and other hybrid forms. These places of work are increasingly utilised by freelance and entrepreneurial creative workers. Bilandzic et al. map Brisbane's CCE ecosystem over time and geographical distribution, and visualise relations among CCEs in terms of purpose, spatial configurations and business entity types. Consistent with global trends, since 2000 the authors found regularly emerging CCEs, dense concentrations of CCEs in the inner city and high numbers of privately owned CCEs. Bilandzic et al. differentiate between *fixed* spaces (usually with standardised, hireable facilities and equipment), *flexible* spaces (with facilities and spaces configurable by users) and *free* spaces (e.g., pop-ups in empty spaces that are user designed). The purposes of the spaces studied are diverse and comprise education and learning; tinkering; art and design production; meetup and networking; independent co-working; art and design exhibition; and retail. Emphasising that economic activity draws value from its social context, the authors conceive Brisbane CCEs as a diverse ecosystem that includes social precursors to commercial activity as well as social innovations themselves. Despite the challenge of a lack of diversity in terms of gender and innovation intent (e.g., profit, art, social good), a well-managed and diverse ecosystem of CCEs can provide creative workspaces with flexible cost and use options, social support for otherwise isolated individual creatives, and connection to work and partnership opportunities.

Thompson continues the focus on footloose freelance creative labour in Chapter 10. Her analysis is based on in-depth interviews with *digital nomads* across Europe, and is informed by critical theorists such as Bauman (2000). In the face of poor employment prospects, millennial digital nomads are attracted by an *imagined* lifestyle of freedom, global travel and entrepreneurial opportunity. Thompson argues that the reality they face is precarious work, social isolation and controlling algorithmic management. She suggests that the digital nomad is perfectly adapted for Bauman's (2000) *liquid times*: an era marked by uncertainty, smaller government, expanding corporate power and fractured work identities. Overall, she concludes that contrary to the "freedom" that the business community attaches to the concept of gig-work, the work intensification demanded by the low wages and increasing platform fees creates a "new global underclass" that is anything but free. Thompson argues that for digital nomads to improve their job security, their summits should exploit their collectivity and advocate for new government regulation in the fight for better worker protection on digital platforms.

In Chapter 11, through an analysis of the popular video-sharing app TikTok, Collie and Wilson-Barnao provide another example of disaggregation and re-distribution of spaces for creative work. The authors argue that while digital technologies and platforms such as TikTok have seemingly "democratised" systems of cultural production and distribution – allowing anyone to produce, curate and share creative content – they have also embedded creative practices and their value within the data-driven logic of the digital economy, with ever more granular surveillance and data capture. Collie and Wilson-Barnao argue that TikTok is remaking vernacular culture, play and creativity into a kind of immaterial and unpaid *digital labour* (Terranova, 2000) – in fact, digital *child* labour, which operates as an exploitative *digital enclosure* (Andrejevic, 2007) with algorithms as the organising force. TikTok exemplifies delegation of the work of "sorting, classifying and hierarchizing of people, places, objects and ideas" (Striphas, 2015, p. 395) to carefully engineered and commercially orientated algorithms. Collie and Wilson-Barnao see TikTok as a portent of a future in which creative labour is increasingly valued for its capacity to generate viral engagement, and produce user data, rather than its particular cultural or aesthetic content. Datafication of creativity enables embodied capital to be turned into separable capital (cf. Bowman and Swart, 2007), disarticulating the value of creativity from the particular bodies and communities where it was produced. This, Collie and Wilson-Barnao suggest, has significant implications for future creative work practices and who profits from the value that is created.

Creatives who work outside the creative sectors are a growing segment of creative work. In Chapter 12, Bowman and Swart conceptualise the management of this kind of creative work as a problem rooted in *causal ambiguity* – that is, the lack of understanding of cause and effect between this kind of creative work and its links to firm performance. The drivers of causal ambiguity are the tacitness of this creative work, the spatial gap between managers and creative work, the complexity of the creative work, and the time lag between the work and firm outcome. Bowman and Swart argue that outcomes of creative work in these contexts may be critical competitive advantages, but if the dominant managerial approaches are primarily geared to the dominant activity, this may be detrimental for workers and the firm. In terms of the future of creative work, the authors suggest that twentieth-century firms often relied on rents from stable, firm-owned resources, and these firms had average life spans of decades. This life span has substantially shortened in the twenty-first century. Stable creative resource rents are not as significant as they once were. This is driven by the rapid diffusion of valuable knowledge across competing firms through the internet, social media, globalisation and a mobile workforce. The extent to which such valuable employee know-how can be partially captured through codification determines the extent to which knowledge shifts from an embodied form of capital to a separable form (Bowman and Swart, 2007). There are two possible alternatives to consider here. In the first, embodied human capital – that is, experience-rich knowledge and skills – remains at the heart of a firm's value offering and digital affordances merely act to create efficiencies. This increases the bargaining power of embedded creatives, particularly in fast-paced, highly connected and competitive labour ecosystems. In the second scenario, the introduction of increasingly sophisticated systems that reduce causal ambiguity for managers replaces the work of traditional embedded creatives, increases separable capital for the firm and reduces the bargaining power of creatives.

EDUCATING FOR THE FUTURE OF CREATIVE WORK

In Chapter 13, Bilton suggests a duality in the definition of creativity, which recognises both novelty and value, and is reflected through different stages in the creative process. In today's creative economy, novel ideas must be made both meaningful and valuable, and co-created with consumers, often using digital channels. This has implications not just for the primary authors of creative works and content, but also for the many ancillary workers who adapt and innovate the systems that make production and distribution possible. Creativity 2.0 describes a networked approach to creativity that integrates artistic imagination with the digital technologies needed to deliver it. The next generation of creative workers will need to cooperate in multi-disciplinary teams, integrating divergent skills. This requires better integration of the creative arts with science and technology disciplines, and building skills for creative work as a collective practice that is aware of diversity and its combinatorial power. In addition to unleashing individual talent, participatory educational ecosystems of collaboration need to be cultivated in educational institutions. Robinson's report on cultural and creative education in the United Kingdom (NACCCE, 1999) is presented as a prescient attempt to address these challenges, advocating a holistic approach to education and skills that extends beyond the school curriculum. However, the failure to implement this highlights the political and pragmatic challenges that confront Creativity 2.0.

Similarly, in Chapter 14, Rodrigues argues that individuals improve their employability when they have deep disciplinary expertise in one creative domain that is connected to additional expertise in multiple alternative domains. This is especially so given that future creative workers may find employment using their creative skills in different sectors from those they were trained for. Rodrigues focuses on how such multi-domain expertise is acquired. This issue is addressed using the innovative *ecological dynamics, constraints-led approach* and is illustrated with a study of the acquisition of dance pedagogical expertise by dancers. Five constraints were identified: (a) individual needs; (b) task rules; and contextual issues related to (c) mentors, (d) peers and (e) students. It is through reaction to constraints that learning occurs. The constraints-led approach suggests that creatives can gain new expertise through:

- formal and informal interaction with mentors and peers;
- engaging in learning environments, with different task constraints, to allow natural evolution of expertise;
- engaging in *dynamic* environments to learn to deal with the unexpected;
- exploring their own intrinsic motivation and goal-setting;
- avoiding the pursuit of a "one right answer" ideal expertise model.

In the concluding chapter, Bridgstock, Tytler and White ask the question: Do creative skills future proof your job? They discuss numerous studies published in the last five years that document escalating changes to the world of work under the influence of digital technologies such as automation and machine learning. Drawing in part on their team's study "100 Jobs of the Future" (Tytler et al., 2019), which was based on interviews with research and technology development experts, this chapter explores research-based predictions about how work and careers will unfold, which capabilities will be required, and the ways in which different kinds of human creativity will be valued. Creative industries roles included various kinds of designers (of autonomous vehicles, 3D printed buildings, gamification, augmented reality experiences), marketers (personalised marketing), content creators (personal brand content) and artists (e.g., swarm artists, who use swarms of hundreds of drones moving in formation to create art, music or performance-based cultural experiences). Bridgstock et al. argue that creativity seems to be an important ingredient in the future of work – both as a capability set that may be "uniquely human" and not easily automatable, and as a key way to add value across different work scenarios. This ingredient allows us to manage and solve complex problems through processes of induction and synthesis.

NOTE

1. ORCID iD: 0000-0003-2245-3433. Creative Industries Faculty, Queensland University of Technology, g.hearn@qut.edu.au.

REFERENCES

Andrejevic, M. (2007). Surveillance in the digital enclosure. *The Communication Review*, **10**(4), 295–317.

Bauman, Z. (2000). *Liquid modernity*. New York, NY: Polity.

Bowman, C. and Swart, J. (2007). Whose human capital? The challenge of value capture when capital is embedded. *Journal of Management Studies*, **44**(4), 488–505.

Haskell, J. and Westlake, S. (2018). *Capitalism without capital: The rise of the intangible economy*. Princeton, NJ: Princeton University Press.

Hearn, G. and McCutcheon, M. (2020). The creative economy: the rise and risks of intangible capital and the future of creative work. In G. Hearn (Ed.), *The future of creative work: Creativity and digital disruption*. Cheltenham, UK and Northampton, MA, USA: Edward Elgar Publishing, pp. 14–33.

NACCCE (1999). *All our futures: Creativity, culture and education*. Report to the Secretary of State for Education and Employment and the Secretary of State for Culture, Media and Sport. Retrieved from http://sirkenrobinson.com/pdf/allourfutures.pdf.

Striphas, T. (2015). Algorithmic culture. *European Journal of Cultural Studies*, **18**(4–5), 395–412.

Terranova, T. (2000). Free labour: Producing culture for the digital economy. *Social Text*, **18**(2), 33–58.

Tytler, R., Bridgstock, R.S., White, P., Mather, D., McCandless, T. and Grant-Iramu, M. (2019). *100 jobs of the future*. Ford Australia. Retrieved from https://100jobsofthefuture.com/report/.

PART I

The evolution of creative work

2. The creative economy: the rise and risks of intangible capital and the future of creative work[1]

Greg Hearn[2] and Marion McCutcheon[3]

INTRODUCTION

The idea of a *creative economy* is a highly contested term, but it has wide international currency and is central to this book. The purpose of this chapter is to articulate what is meant by the term *creative economy*, and to explain the way it can be operationalised methodologically. Criticisms of this methodology are discussed and responded to. It is then argued that the rise in intangible investments, globally, parallels the growth in the creative economy because intangible assets require creative innovations that are rare and hard to replicate in order to confer competitive advantage. The correspondence between types of intangible assets and categories of creative occupations is described. The characteristics of intangible assets that are different from tangible assets are also described. Finally, normative issues in creative economy work are discussed, including the potential shift in axes of disadvantage for creative workers.

The empirical basis for the definition of creative economy in this chapter rests on work carried out by the Centre for Creative Industries and Creative Innovation by Cunningham and others (e.g., Cunningham, 2011; 2014; Higgs and Cunningham, 2007). The empirical core of this research, based on national census occupational data, has been replicated across census periods (2001 to 2016 in Australia), in other countries, and in city and regional jurisdictions. This data is based on individuals' qualitative descriptions of their primary source of work in their census return, which is then coded into national standardised categories of occupation. Each individual is then also coded as belonging to a standardised industry sector derived from their place of work.

In this research approach, the aggregations of *industry segments* that are used to define the cultural/creative industries are:

- advertising and marketing;
- architecture and design;
- software and digital content;
- film, television and radio;
- music and performing arts;
- publishing;
- visual arts.

Aggregations of creative *occupations* also use these same descriptors. The theoretical model then segments occupations into two groups: creative services occupations (advertising and marketing; architecture and design; and software and digital content) and cultural production occupations (film, television and radio; music and performing arts; publishing; and visual arts) (Cunningham, 2014). By cross-tabulating occupation with industry, creative occupations can be identified and enumerated in the economy, not only in terms of *specialists* (those employed in creative occupations in creative/cultural sectors), but also in terms of *embedded* creatives (those employed in the same creative occupations in all other industries) (e.g., Cunningham and Higgs, 2009). Using these statistical definitions, the Australian Bureau of Statistics (ABS, 2014), in its inaugural *Cultural and Creative Activity Satellite Accounts*, developed Australia's "first experimental satellite accounts measuring the economic contribution of cultural and creative activity in Australia" ("Preface", para. 2). The Australian Bureau of Communications and Arts Research (BCAR, 2019) used this statistical framework to analyse creative skills and human capital in the creative economy in Australia. The Regional Australia Institute also uses this schema. Nesta has utilised and further developed these frameworks using United Kingdom (UK) standard occupation codes (SOC) data (Bakhshi et al., 2015).

Beyond simple enumeration, what is the theoretical significance of counting occupations and how is it relevant to understanding the creative economy? Hearn (2014) argues that "an occupation is first and foremost a set of knowledge practices, or more expansively, knowledge, skills, identity formations, social relationships and practices" that have "been subjected to evolutionary processes through the labour market and the economy of firms" (p. 90). As will be argued in this chapter, supply and demand for knowledge enacted in occupations (i.e., doing a job with a certain skill set) is a key to understanding the rise of intangible investments because these investments take the form of new knowledge, novel creative affordances and hard-to-replicate relationships.

We will argue that understanding the link between intangible investments and creative work is important to analysing the future of creative work.

While census data is accepted by most researchers as a robust accounting of primary occupational work, this overall *creative economy* research framework has been criticised in the following main ways:

1. It ignores other modes of work: secondary occupations and volunteer or other unpaid work.
2. Some statistical aggregations, particularly *software and digital content*, include statistical sub-categories that should be excluded.
3. Some categories, for example *advertising and marketing* and *software and digital content*, are not creative cultural work.
4. Creative cultural occupations in other industries do not involve "real" cultural/creative work.
5. There is no theoretical basis for connecting all these categories of work into a coherent construct called the creative economy because there is no necessary and "natural" connection between the sectors described above. For example, *music and performing arts* has no relationship (in terms of work modes, practices or economic connection) to *advertising and marketing*.

In terms of statistical critiques, it is true that secondary and other forms of work are not included and other approaches are needed to address them (e.g., Bennett et al., 2014). And in many developing countries, there are simply no formal occupation statistics. This does not make the arguments in relation to primary occupation invalid. It is also true that the statistical category *software and digital content* is perhaps over-inclusive, but it arises because it is difficult to statistically disaggregate some current computer-related occupational codes into non-creative and creative (QUT, n.d.-a).

For example, the games industries incorporate many occupations coded into the general category of information technology (IT) professions. However, of the 21 occupations in this category, the creative economy framework counts only eight occupations, excluding database, systems administration, security, network, support, systems testing, and telecommunications engineering occupations (QUT, n.d.-a). The included occupations focus on multimedia and web developers, business and systems analysts, and software developers. To this is added graphic, web and multimedia designers and illustrators from the design occupation grouping. As computing continues to evolve towards visual and highly interactive interfaces, there is increasing argument for the current categorisation. And in terms of the future of work, the addition of humanoid mechatronic agents and chatbots using artificial intelligence and machine learning will make this disaggregation even more difficult.

In terms of the validity, coherence and interconnectedness of the taxonomy of the creative economy, the first critical problem is semantic: specifically, the imprecision of the word *creative*. Creativity clearly includes scientific production of novelty, for example, and this should not be included in an accounting of what most would consider *creative and cultural work*. However, as it is used in the model under discussion, creative is actually a catchall to mean *replicative or novel aesthetic and/or expressive knowledge, either separately or in synthesis with other forms of knowledge*. Under this definition, it is hard to say that advertising and marketing, for example, are not creative, and have nothing to do with music and the performing arts. Moreover, it is an important corrective to much innovation policy and practice that ignores aesthetic and expressive knowledge altogether.

The second problem with the coherence critique resides in as yet unresolved normative debates about the ontology of work, public and private value, and corresponding policy corollaries and political praxis. In other words, the critique is normative (what ought to be) rather than positivist (what can be counted). However, both critical and positivist scholars agree that the *culturalisation of the economy* is real. Grand theory critiques of capitalism recognised the existence of the creative economy, albeit in critical terms, long before mainstream economics measured it (e.g., Harvey, 1989; Jameson, 1991). Scott (2004) deftly suggests: "One of the peculiarities of modern capitalism is that the cultural economy continues to expand...as an expression of the incursions of sign-value into ever-widening spheres of productive activity as firms seek to intensify the design content and styling of their outputs in the endless search for competitive advantage" (pp. 462–3). This is the unifying dynamic that gives credence to the idea of the creative economy, both for empirical and critical purposes.

Finally, that the creative economy as empirically defined above is a coherent and measurable construct can be inferred from standard national economic reporting. Australia is not an atypical modern economy and ranks 14th in the world in GDP. Australia serves here as a proxy model for all developed economies. For example, the interpenetration of creative work into other sectors can be inferred from national educational qualification data (BCAR, 2019). Some sectors employ people with creative skill qualifications more intensively than others. For example, in Australia, the information media and telecommunications sector employs the highest number of people with creative qualifications: 27 per cent of its workforce. At the other end of the scale, about 3 per cent of workers in the agriculture and mining sectors have creative qualifications. Three of the fastest growing sectors in Australia – professional, scientific and technical services; rental, hiring and real estate services; and information media and telecommunication services – employ the highest share of workers with creative qualifications, suggesting that this human capital has actually been important in accounting for part of their growth (BCAR, 2019).

Standard economic input and output analyses also show how different industries take inputs from creative industries as intermediate inputs in producing their final products and services. Professional services, computer services, insurance, public administration, finance and retail are all heavy users of creative inputs, in addition to employing large numbers of creatively qualified people. For example, *professional scientific and technical services* utilises AU$12 billion worth of creative inputs annually, and *computer systems, insurance and superannuation*, along with *public administration and regulatory services* and *finance*, each use around AU$6 billion of creative inputs (BCAR, 2019). In total, creative industries produced goods or services to the value of AU$87 billion that were used by companies in other sectors in Australia in 2017 (BCAR, 2019). In Australia, there are more creative occupations outside of the core creative industries than inside. Roughly a little over half of all creative jobs are in other industry sectors. In 2016, there were 185 000 formally classified creative workers embedded in other industries compared to 162 000 in creative industries companies (QUT, n.d.-b). The annual employment growth between 2011 and 2016 for embedded creative workers in other industries is 2.9 per cent, which is 2.4 times higher than the average of the rest of the Australian workforce (QUT, n.d.-b). This rate is also higher than the average growth of specialist creative jobs in creative companies, which is 2.1 per cent per year, and 1.75 times higher than national average job growth (QUT, n.d.-b).

We now turn to this question: Is there a connection between the rise of intangible capital and the growth of cultural and creative jobs across the whole economy described above? In what follows, the connection between intangible capital and creative work is first described and then normative issues for creative work are discussed.

THE RISE OF INTANGIBLE CAPITAL AND CREATIVE WORK

In the mid-1990s, United States businesses for the first time invested more money in intangible assets than tangible assets, and the same thing occurred in the UK in the early 2000s (Haskell and Westlake, 2018). Moreover, this watershed moment of higher investment in intangibles has continued. But what is meant by an *intangible asset*? Assets are enduring, economically productive, and can be owned. Tangible assets are familiar to most (e.g., buildings and equipment). Haskell and Westlake (2018) suggest there are three broad categories of intangible asset:

1. investments in computerised information (primarily software and data);
2. investments in innovations such as original creative works, other intellectual property (IP), research and development (R&D) and designs;
3. investments in economic competencies (e.g., assets used in marketing and branding, or increasing organisational competence).

It is important to note that much creative work is not concerned with the creation of *enduring* creative assets, but nevertheless still creates intangible value, for example, in day-to-day communications, creative performances, or aesthetic flair in interior design. However, many of the outputs of creative work do produce enduring assets that can be protected as IP. Furthermore, there is a parallel between these intangible asset classes and categories of creative occupations, for example:

1. Computerised information assets may require software and digital content occupations.
2. Original creative works, R&D, and designs may require digital content, cultural production or design occupations.
3. Marketing and business assets may require advertising, design and communications occupations.

Creative work could arguably be involved in all three categories of intangible asset development to different degrees. Software development increasingly has important aesthetic and user-experience elements, which creative workers specialise in (e.g., virtual creating, virtual worlds, augmented reality, digital communication and web design). Design of ownable products and services is important for innovation leadership at the start of the supply chain (Mudambi, 2008). Market research and branding assets enable control of the consumption end of the supply and may require aesthetic digital or user-experience input (Mudambi, 2008). Training and business assets may also utilise creative workers because these activities are intended to affect the behaviour of people and therefore involve communication and other forms of creative engagement. One novel example is the use of creative arts approaches to the design of more engaging and effective experiential business processes and training outcomes (Bilton and Cummings, 2014). A notable commercial example of the utilisation of creative competencies is IDEO, which invented a new class of consultancy around the concept of *design thinking*.

Given the long-term trend towards increasing investment in intangible assets and growth in many categories of creative work, the key question to be asked is why would businesses invest in these intangibles? There are a number of reasons for this investment, but all are underpinned by a belief that these investments are important to the continued success of the company in question. And the longevity of the trends towards that investment would suggest that they *may be important* to business success. The general argument here is that companies survive if they can more competitively create value to be consumed either by other businesses or private individuals. The amount of profit that is generated is proportional to the extent to which the company can out-compete those that would seek to disrupt their market position. The resource-based view of the firm (Barney, 1991; Peteraf, 1993) suggests that companies that can deploy rare and hard-to-duplicate resources are able to produce and capture more value. This phenomenon has been used to explain performance in design consultancies (e.g., Abecassis-Moedas et al., 2012) and in Hollywood (Miller and Shamsie, 1996). This theory is one explanation of the growth of the embedded creative workforce (Hearn and Bridgstock, 2014).

Most recently, Haskell and Westlake (2018) have provided an extensive explanation for the growth in intangible investments by companies. They suggest there are four characteristics of intangible investments that give them distinctive and different economic properties to tangible investments and confer sometimes dramatic competitive advantage. Haskell and Westlake summarise these as the four S's: scalability, sunkenness, spillovers and synergies. These characteristics are discussed and exemplified below.

Scalability

Physical assets are typically excludable and cannot be shared, or duplicated without additional cost. However, ideas and concepts in the form of knowledge can be used over and over again by different entities in different places. So scalability has got to do with the ease with which a business can reproduce an intangible asset without additional fixed costs (or limited additional cost) to increase the scale of their operation. A simple example of scalability is to be found in the comparison of food with the recipe. Food can be consumed by only one person and has to be transported to different consumers; however, a recipe, once created, can be shared infinitely, particularly in the era of the internet, at little or no cost. A related aspect of scalability is the idea of *network externalities*. A network externality exists when an asset increases in functional or perceived value in proportion to the number of other instances of it that are connected. For example, mobile phones are more functionally valuable when more people have them; however, a table's value is not affected when more people have one. Visa card, Facebook and Uber are companies whose value is built on high network externalities, rather than factories, land or equipment.

The foundational analysis of scalability was provided by Arthur (1996), who explained why early success in markets, underpinned by scalable, networked technologies, leads to a cycle of increasing returns to those companies that gain the lead with early market success. Arthur explained that in contrast to, for example, extractive industries in which there are diminishing returns as the resource is depleted, software companies, for example Microsoft, benefit from network effects due to the requirement or advantage of compatibility between systems, as well as *customer groove in*, which occurs because of relative high levels of time and skill required to learn a system and to change it. Once sufficient numbers of users are *locked in* to a software platform, the market evolves to a *winner takes all* dynamic. Even today, with other options available, there are 900 million users of Windows 10. According to Arthur, even industries with a primarily tangible product and assets, for example fast food, can exhibit increasing returns through franchising and branding.

Sunkenness

Another way of saying this is high upfront costs and low distribution costs. A sunk cost is an investment that is required to create a product or service, but that cannot be recouped easily in the event of failure (Haskell and Westlake, 2018). Tangible assets can be resold if a business fails. Buildings retain their value, at least to a large extent, despite the fact that a particular company may not be able to use them profitably. On the other hand, intangible assets are more difficult to resell in the event of failure; for example, advertising campaigns that create a brand for a company in its start-up phase are a sunk cost that is very difficult to recoup any value from once that business has failed. Equally, as Arthur (1996) notes, higher upfront costs are a deterrent to smaller entrants. This sunkenness of intangible assets creates a very distinctive set of financial system corollaries in terms of investment, financial tools and risk management that has had enabling and constraining effects on the growth of the creative economy (Haskell and Westlake, 2018).

Spillovers

A spillover in economic terms is when one person's investment creates advantages for other investors. If someone buys a warehouse, this may have benefits for neighbours, but these would be indirect benefits, such as strengthening the industrial ecosystem. Neighbours can only gain a direct benefit if they pay for a sub-lease, or break in and steal something. With intangible investments, copying ideas is relatively easy to do. Therefore, competitors can directly benefit from innovative ideas that are adopted and easy to copy. Consider the competitors emerging who compete against Uber, such as Didi and Ola. These companies are essentially copying the innovation that was a new product class created by Uber. Patents offer some protection against such spillovers, but there are many kinds of innovative ideas that cannot be copyrighted or patented. Haskell and Westlake (2018) make the point that spillovers are important for two reasons. First, if companies cannot be confident that they will obtain benefits from these intangible investments, they would not be as inclined to invest. Second, there is a premium on the ability to control the spillover problem and companies that can protect their investments in intangibles are more likely to be profitable.

Synergies

Intangible investments often enhance their innovativeness by bringing together unique combinations or innovations. Characteristic of innovation capital is that it is often combinatorial. In other words, more unique innovation comes from a combination of new ideas and technologies, rather than one idea itself (Haskell and Westlake, 2018). It can be argued that innovative technologies can come from any of the four modern *knowledge paradigms*: science and technology, economics and business, creative expressive, and social (e.g., Hearn and Rooney, 2008). These knowledge paradigms roughly map to clusters of disciplines and associated knowledge work: STEM (i.e., science, engineering and technology), business (i.e., economics, finance, entrepreneurship and management), social sciences (sociology, psychology, policy, law and social work) and MAD (media, arts and design).

Synergistic innovations within any one of the fields are easier to achieve because they operate from similar positions on ontology and epistemology. Similarly, synergies between two *adjacent*[4] fields are somewhat easier, whereas synergies between fields that are very different are harder to achieve. More importantly, innovations that incorporate synergies from all four knowledge paradigms are more difficult still. What this means in practice, for example, is that a new product or service that incorporates technology innovation, business-model innovation, social innovation and novel design is hard to replicate.

As a result, it is very rare to find companies that excel in all four knowledge paradigms. EMI is a company that excelled not only in technology, but also in the creative arts and in the execution of business (cf. Haskell and Westlake, 2018, p. 59). The entertainment sector incorporates all four paradigms to different degrees. Apple excels at technology, business, design and creative content. Some platform innovations, for example virtual reality, have already broken down boundary conditions between the different knowledge paradigms. The computer games industry is the best example of this, which required the coming together of very strong technical, narrative and creative capabilities. Hollywood itself has always been at the forefront of technology innovation and rapid adoption.

So far in this chapter, the intent has been to:

- define what is meant by the term *creative economy* and empirically specify it;
- explain the rise and relevance of intangible investments to the creative economy;
- suggest why companies make these investments in intangible capital;
- explain the relationship of intangible assets to the continued growth in creative occupations right across (developed) economies.

The normative questions that are raised by these trends in creative work are now considered.

NORMATIVE ISSUES IN THE FUTURE OF CREATIVE WORK

Normative critiques of academic and policy discourse that use the terms *creative industries* or *creative economy* have been prosecuted from within cultural studies and cultural geography in recent decades (e.g., Oakley and O'Connor, 2015). The key tenet of relevance here concerns the precarity of creative labour. For example, in their comprehensive critique of creative industries studies and policy, *The Routledge Companion to the Cultural Industries*, Oakley and O'Connor (2015) point out that "…the creative economy is perfectly compatible with the most egregious forms of exploitation, inequality, and economic disenfranchisement" (p. 5). This is true. However, as Solow (1987) opined, "You can see the computer age everywhere but in the productivity statistics" (p. 36); similarly, in Australia at least, it could be argued that creative labour precarity is everywhere except in national employment statistics. For example, average salaries for those in full-time creative economy work, as defined above, are in fact higher than national salary averages and, as noted above, the number of these full-time jobs is growing up to 2.5 times faster than average job growth. This is true even though the calculation includes artists, who, as a group, are well below the national average. The IT productivity puzzle was eventually solved by closer attention to definition and measurement approaches (see Haskell and Westlake, 2018). It can similarly be argued that more nuanced specification of creative labour contexts is needed to specify precarity among creative workers in various contexts. This would advance normative debate and evidence-based policy, and empower creative-labour advocacy.

As per Table 2.1, between 2011 and 2016, all occupation groups in creative services experienced job growth higher than that of the Australian workforce overall, while positions in all occupation groups in cultural production grew at a slower rate than the Australian workforce, or decreased (QUT, n.d.-c).

Creative services occupations:

1. *Advertising and marketing.* Between 2011 and 2016, the strongest growth in positions was in this sub-sector, which grew by an annual average of 4.7 per cent, nearly 3500 positions per year.
2. *Architecture and design.* This sub-sector experienced the third-largest increase in positions between 2011 and 2016, just over 2000 positions per annum.
3. *Software and digital content.* This group had the second-highest increase of 3.2 per cent and nearly 2400 positions per year.

Table 2.1   Estimates of employment and growth by creative industry occupation, 1996–2016

	Persons			Share of total creative occupations (%)			Average annual growth (%)	
	1996	2011	2016	1996	2011	2016	1996–2016	2011–16
Creative services occupations								
Advertising and marketing	31,530	67,180	84,510	16	22	24	5.1	4.7
Architecture and design	43,860	81,250	91,500	23	27	26	3.7	2.4
Software and digital content	46,780	70,870	82,820	24	23	24	2.9	3.2
Total	122,170	219,300	258,840	63	72	75	3.8	3.4
Cultural production occupations								
Film, television and radio	13,570	21,100	22,420	7	7	6	2.5	1.2
Music and performing arts	13,280	20,250	22,290	7	7	6	2.6	1.9
Publishing	25,020	32,290	30,390	13	11	9	1.0	−1.2
Visual arts	18,730	13,610	13,200	10	4	4	−1.7	−0.6
Total	70,600	87,240	88,290	37	28	25	1.1	0.2
Total creative occupations	192,770	306,540	347,130	100	100	100	3.0	2.5
Total workforce	7,632,690	10,057,150	10,683,840	-	-	-	1.7	1.2

Source: QUT (n.d.-c). Occupational aggregations based on Cunningham (2014).

Cultural production occupations:

1. *Film, television and radio*. Between 2011 and 2016, positions in this occupation group grew at the same rate as the Australian workforce, accumulating just over 250 additional positions each year.
2. *Music and performing arts*. The number of positions in this category grew at a faster rate than the Australian workforce, at an average annual rate of 1.9 per cent, just over 400 positions per annum.
3. *Publishing*. The number of positions recorded in this sub-sector fell between 2011 and 2016, resulting in an average loss of nearly 400 positions each year.
4. *Visual arts*. The number of positions in this sub-sector has fallen since 1996, although the rate of decrease slowed between 2011 and 2016, to an average rate of −0.6 per cent or nearly 100 positions each year.

Table 2.2   *Estimates of mean income by creative occupation segments, 1996–2016 (2016 Australian dollars, adjusted using CPI)*

	Mean incomes			Difference compared with total workforce (%)			Average annual growth (%)	
	1996	2011	2016	1996	2011	2016	1996–2016	2011–16
Creative services occupations								
Advertising and marketing	$61,400	$80,100	$81,600	32	36	33	1.4	0.4
Architecture and design	$55,400	$65,400	$66,300	19	11	8	0.9	0.3
Software and digital content	$78,500	$91,800	$93,100	68	56	52	0.9	0.3
Total	$65,800	$78,500	$79,900	41	33	30	1.0	0.4
Cultural production occupations								
Film, television and radio	$69,000	$75,100	$74,500	48	27	21	0.4	−0.2
Music and performing arts	$40,200	$45,100	$44,200	−14	−24	−28	0.5	−0.4
Publishing	$57,800	$65,400	$66,400	24	11	8	0.7	0.3
Visual arts	$36,300	$43,700	$45,700	−22	−26	−26	1.2	0.9
Total	$50,900	$59,600	$59,700	9	1	−3	0.8	0.0
Total creative occupations	$60,400	$73,100	$74,800	30	24	22	1.1	0.5
Total workforce	$46,600	$59,000	$61,400	-	-	-	1.4	0.8

Source: QUT (n.d.-c). Occupational aggregations based on Cunningham (2014).

As per Table 2.2, creative industry occupation mean incomes are relatively stable compared both in relation to each other and the total Australian workforce, with each occupation group generally maintaining its income ranking over time. In nearly all occupation categories, creative industry incomes are on average higher than that of the Australian workforce, but their growth has not kept pace with growth in total workforce income (QUT, n.d.-c).

Creative services occupations:

1. *Advertising and marketing*. This group earned the second-highest creative industry mean income, with strong growth in positions suggesting demand for personnel is supporting wages. Five-year income growth is half that of the total workforce, following strong growth up to 2011.
2. *Architecture and design*. Incomes here are growing more slowly than in other creative industry occupations, with the mean converging on that of the national workforce.
3. *Software and digital content*. Since 1996, this occupation group has consistently earned the highest creative industry mean income. Although growth in the number of positions in this category exceeded that of the total workforce, income growth slowed in the five years to 2016, at one quarter of that of the workforce.

Cultural production occupations:

1. *Film, television and radio*. Mean incomes here fell in real terms between 2011 and 2016, decreasing by an average of 0.2 per cent per annum. This occupation category is the only one to have fallen in its ranking by income, from second in 1996 to third in 2011 and 2016.
2. *Music and performing arts*. This is the lowest-earning creative occupation. Mean income fell by an average of 0.4 per cent per annum, while the workforce in this category continued to grow at a rate faster than the total workforce.
3. *Publishing*. The mean income in this sub-sector is growing at about half the rate of that of the total workforce. The number of positions in the sub-sector is falling at a faster rate than in any other creative industry category.
4. *Visual arts*. Although this sub-sector is one of the lowest-income categories and has a shrinking workforce, between 2011 and 2016 mean income in the sub-sector grew at a faster rate than that of the Australian workforce.

From Tables 2.1 and 2.2, there are number of key observations to be made in terms of full-time employment and the salary conditions of creative workers:

• Total wages growth in Australia between 2011 and 2016 was weak at just 0.8 per cent per annum.

- In general, creative sector employment opportunities and salaries are superior to those of the average worker and far better than some sectors. For example, employment in manufacturing in Australia declined between 2006 and 2016.
- The creative occupations hardest hit in terms of employment are visual artists and those categorised in publishing, specifically authors, journalists, archivists, curators and librarians.
- Even though salaries in music and the performing arts are lower than the national average, employment numbers show relative strength compared to the economy as a whole and other categories of creative occupation. This suggests that some creative occupations have compensating intrinsic benefits apart from wages (cf. Potts and Shehadeh, 2014).

Taken as a whole, full-time work in the creative economy does not indicate precarious conditions when compared to the rest of the Australian workforce. However, consistent with most industry sectors, creative work is likely to be more precarious for those in part-time and casual work. This is more likely to be so in contexts that operate outside formal labour regulatory institutions, for women, youth and other minorities, and for those in developing countries who are part of global supply chains, or excluded from them. In addition, different contexts of full-time work need to be considered (e.g., national, sectoral, and urban versus regional). Also, for full-time work, relative inequality continues to be an issue. For example, the overall gender pay gap for Australian creative workers is around 25 per cent[5] (ABS, 2019). This pay gap is in the highest quartile of gender-based pay gaps in Australia, a quartile that primarily involves high-skilled knowledge occupations (Workplace Gender Equality Agency, 2019). Although salaries are high compared to the rest of the work force, the gender pay gap is a significant issue. Moreover, the gender pay gap for creative workers increases to around 33 per cent in regions outside the great capital cities (ABS, 2019).

Some have argued that the shift in capital investment from tangibles to intangibles will create new axes of disadvantage for creative workers, or exacerbate existing ones. For example, Haskell and Westlake (2018) provide evidence that inequality between locales is more significant than inequality within locales. Cities beat regions because cities promote more spillovers and synergies (Haskell and Westlake, 2018). Wealth inequality, based on housing equity, is also more pronounced between cities and regions because land is scarcer in cities and those with intangible-economy jobs earn more and can pay more (Haskell and Westlake, 2018). Similarly, inequality within firms that have intensive intangible capabilities is less pronounced than it is for firms that do not have these capabilities. Cities with diverse economies do better than those anchored in one sector. Because creative work is concentrated in urban contexts, creative workers are likely to experience far less precarity than, for example, unskilled workers in non-urban contexts.

Durand and Milberg (2019) describe the role of intangibles in the operation of monopolies in global value chains. Within global product development and supply, intangible investments are made in the initial R&D/design stage, as well as the final marketing/advertising and customer relationships (Mudambi, 2008). Durand and Milberg (2019) suggest that these global value chains feature asymmetries: those with intangible investments capture a bigger share of profit than those with primarily tangible investments (e.g., manufacturing). That is, "the capture of value added is largely detached from the flow of physical goods and mainly related to intangible aspects of the supply chain" (Durand and Milberg, 2019, p. 405). These monopolies capture and control profits at the beginning and end of the supply chain (in developed countries), while simultaneously pushing down costs, conditions and pay in the manufacturing and production "middles" of the supply chain. However, relatively speaking, manufacturing labourers in developing countries have done better than those in developed countries in terms of wages growth. In addition, this has destabilised the competitive landscape between firms. The retail war between Walmart and Amazon has been a war between tangibles versus intangibles, though Walmart has increasingly ramped up intangible investments, for example, in terms of advanced consumer data analytics (Durand and Milberg, 2019). Arguably, this behaviour will have flow-on effects in the labour market that is servicing Walmart and in competition between occupational categories. Because intangible capital can be more easily transferred between national jurisdictions than tangible capital, companies can easily optimise their tax obligations in their favour. This lessens national tax revenues and hence labour regulation enforcement in some jurisdictions.

Durand and Milberg (2019) suggest a taxonomy of *rents* related to overlapping intangible investment characteristics such as IP rights, network externalities and intangibility scalability. The four kinds of rents are legal IP rent, vertical natural monopolies rent, intangibles differential rent, and data-driven innovation rent. Legal IP rent derives mainly from patents and copyright (e.g., Louis Vuitton); vertical natural monopoly rent arises from vertically integrated supply chains, such as Apple, and all of the legal and organisational intangibles that allow virtually unilateral control of the global value chain. The best example of intangibles differential rent is Nespresso versus coffee producers. Data-driven innovation rents are exemplified by, for example, Amazon's analytics on shopping histories. The taxonomy suggests differentiation in power and work contexts and required regulation, which future research needs to analyse "within the complex entanglement of international vertical interdependencies that characterize [global values chains] today" (Durand and Milberg, 2019, p. 423).

Apart from the implications of economic change resulting from increasing investments in intangible capabilities, there is an even more basic disruption of social power that arises in the global digital networks that enable intangible assets to be deployed. The intangible economy is built on digital networks that connect components of the supply chain, and consumers to corporations, anywhere in the world. However, digital networks also connect workers to workers, workers to jobs (anywhere in the world), and citizens to citizens. This is an unprecedented evolutionary step-change in *social structure*: the "distinctive, stable arrangement of institutions whereby human beings in a society interact and live together" (Wilterdink and Form, 2018, para. 1). This has resulted in a new and complex shift in agency in power relations. Network science has investigated the structural properties of such networks (as opposed to hierarchies) for some time. Barabási (2014) exposited that digital networks evolve according to *scale-free patterns* that are found in many other phenomena – for example, the internet, social networks, citation networks, protein structure, semantic networks, trade between cities, and Hollywood actors. All scale-free networks are characterised by:

- dynamic evolution;
- no complete central control;
- hub nodes with very large numbers of connections and intermediate modular nodes with above-average numbers of connections (although most nodes have a smaller number of connections);
- hubs that offer *preferential attachment properties* with increasing returns to scale (cf. Arthur, 1996);
- exogenous preferential attachment properties due to scale advantage through network externalities, which are 1000 per cent more powerful than the original endogenous properties that caused the hub to grow (Arthur, 1996);
- evolution of the number of connections per node towards a power curve rather than a normal curve;
- strong resistance to failure by hubs.

In the hierarchies that were found in the companies and social institutions of the twentieth century, social power was concentrated in a few people. Defeating that power required workers to form similarly large hierarchical structures that were similarly powerful. In the digital economy of the twenty-first century, social power operates differently, not necessarily more democratically, and definitely not usually with egalitarian outcomes. For a start, most organisations in the network are still hierarchies. And several hierarchical hub companies exert one-sided control over networks of contract workers (e.g., Uber), consumers (e.g., Amazon) or citizens (e.g., Facebook), supply chain subcontracting partners (e.g., Apple), or the supply of work (e.g., Airtasker). At the same time, digital content workers can connect to markets anywhere in the world, and independent artists have a platform to operate outside of traditional corporate IP regimes and artist control. Social activism can be more easily crowdfunded. Individuals can challenge abuse by the most powerful and corporations live in fear of social media contagion destroying brand equity. Powerful twentieth-century corporate incumbents have been routinely dislodged. In addition, there are new forms of collective labour mobilisation around the world (Lazar and Sanchez, 2019).

CONCLUSIONS

Critical and empirical scholars recognise the culturalisation of the economy. Debate exists about whether creative services work and creative occupations in other industries do involve "real" cultural/creative work and whether there is a theoretical basis for connecting all these categories of work into a coherent construct called the creative economy. In this chapter, creative work is defined as that which involves *replicative or novel aesthetic and/or expressive knowledge, either separately or in synthesis with other forms of knowledge.* It is argued that the creative workforce in the creative economy extends beyond traditional arts, media and cultural occupations and sectors, and can be specified and measured in terms of national occupation statistics that are consistent with this definition. The coherence of the creative economy as a construct is supported by the accelerating digitisation of all aspects of creative/cultural work and demonstrable economic interdependencies right across the economy. The growth of the creative economy workforce is paralleled by the rise of intangible investments in assets derived from human capital. Categories of intangible assets have a close correspondence with creative economy occupations. The rise in intangible investment creates opportunities for the creative economy workforce, but also disrupts power relations in the world of work, potentially leading to new axes of disadvantage for the creative workforce. For full-time workers in the creative economy, in developed economies, precarity exists, but is less severe than for many other workers. Casual and part-time creative workers are likely to experience real precarity and relative inequality. Precarity and inequality as creative workforce issues are particularly significant for women, those outside large cities and those in traditional arts-related roles.

NOTES

1.   This chapter was completed with funding from the Australian Research Council Linkage project (LP160101724) led by Queensland University of Technology in partnership with the University of Newcastle, Arts Queensland, Create NSW,

Creative Victoria, Arts South Australia and the WA Department of Culture and the Arts. The chapter uses excerpts from QUT (n.d.-c). The views expressed are those of the authors alone.

2. ORCID iD: 0000-0003-2245-3433. Creative Industries Faculty, Queensland University of Technology, g.hearn@qut.edu.au.
3. ORCID iD: 0000-0003-3755-9070.
4. By adjacent we mean having shared subset assumptions. For example, data-based decisions are shared by both STEM and some aspects of business.
5. The causes of the enduring general gender pay gap are complex. Hedijer (2017) suggests a well-accepted categorisation of causes is between potential life situation variables that are known to predict pay gaps (e.g., age, education, health, relationship status, occupation, size of company, and employee versus managerial role) and what cannot be explained by these variables, which is attributed to direct discrimination.

REFERENCES

Abecassis-Moedas, C., Ben Mahmoud-Jouini, S., Dell'Era, C., Manceau, D. and Verganti, R. (2012). Key resources and internationalization modes of creative knowledge-intensive business services: The case of design consultancies. *Creativity and Innovation Management*, **21**, 315–31.

Arthur, W.B. (1996). Increasing returns and the new world of business. *Harvard Business Review*, **74**(4), 100–109.

Australian Bureau of Statistics (ABS) (2014). *Australian national accounts: Cultural and creative activity satellite accounts* (Cat. no. 5271.0). Retrieved from http://www.ausstats .abs.gov.au/ausstats/subscriber.nsf/0/EFFE2547EC51F5AACA257C78000C1B53/ $File/52710_2008-09.pdf.

Australian Bureau of Statistics (ABS) (2019). *Census of population and housing 2016*. Tablebuilder. Findings based on the use of ABS Tablebuilder data retrieved 2019.

Bakhshi, H., Davies, J., Freeman, A. and Higgs, P. (2015). *The geography of the UK's creative and high-tech economies*. London: Nesta.

Barabási, A. (2014) *Linked*. New York, NY: Basic Books.

Barney, J.B. (1991). Firm resources and sustained competitive advantage. *Journal of Management*, **17**, 99–120.

Bennett, D., Coffey, J., Fitzgerald, S., Petocz, P. and Rainnie, A. (2014). Looking inside the portfolio to understand the work of creative workers: A study of workers in Perth. In G. Hearn, R. Bridgstock, B. Goldsmith and J. Rodgers (eds), *Creative work beyond the creative industries: Innovation, employment and education* (pp. 158–72). Cheltenham, UK and Northampton, MA, USA: Edward Elgar Publishing.

Bilton, C. and Cummings, S. (eds) (2014). *The handbook of creativity management*. Cheltenham, UK and Northampton, MA, USA: Edward Elgar Publishing.

Bureau of Communications and Arts Research (BCAR) (2019). *Creative skills for the creative economy* (Working paper, January). Retrieved from Australian Government Department of Communications and the Arts website: https://www.communications .gov.au/publications/creative-skills-future-economy.

Cunningham, S. (2011). Developments in measuring the "creative" workforce. *Cultural Trends*, **20**(1), 25–40.

Cunningham, S. (2014). Creative labour and its discontents: A reappraisal. In G. Hearn, J. Rodgers, B. Goldsmith and R. Bridgstock (eds), *Creative work beyond the creative industries: Innovation, employment and education* (pp. 25–46). Cheltenham, UK and Northampton, MA, USA: Edward Elgar Publishing.

Cunningham, S.D. and Higgs, P.L. (2009). Measuring creative employment: Implications for innovation policy. *Innovation: Management, Policy and Practice*, **11**(2), 190–200.

Durand, C. and Milberg, W. (2019). Intellectual monopoly in global value chains. *Review of International Political Economy*, **27**(2), 404–29.

Harvey, D. (1989). *The condition of postmodernity: An enquiry into the origins of cultural change*. Oxford, UK: Blackwell.

Haskell, J. and Westlake, S. (2018). *Capitalism without capital: The rise of the intangible economy*. Princeton, NJ, USA and Oxford, UK: Princeton University Press.

Hearn, G. (2014). Creative occupations as knowledge practices: Innovation and precarity in the creative economy. *Journal of Cultural Science*, **7**(1), 83–97.

Hearn, G. and Bridgstock, R. (2014). The curious case of the embedded creative: Managing creative work outside the creative industries. In S. Cummings and C. Bilton (eds), *The handbook of creativity management* (pp. 39–56). Cheltenham, UK and Northampton, MA, USA: Edward Elgar Publishing.

Hearn, G. and Rooney, D. (eds) (2008). *Knowledge policy: Challenges for the 21st century*. Cheltenham, UK and Northampton, MA, USA: Edward Elgar Publishing.

Hedijer, V. (2017). Sector-specific gender pay gap: Evidence from the European Union countries. *Economic Research-Ekonomska Istraživanja*, **30**(1), 804–1819.

Higgs, P. and Cunningham, S. (2007). *Australia's creative economy: Mapping methodologies*. Retrieved from https://eprints.qut.edu.au/6228/.

Jameson, F. (1991). *Postmodernism, or, the cultural logic of late capitalism*. Durham, NC: Duke University Press.

Lazar, S. and Sanchez A. (2019). Understanding labour politics in an age of precarity. *Dialectical Anthropology*, **43**(1), 3–14.

Miller, D. and Shamsie, J. (1996). The resource-based view of the firm in two environments: The Hollywood film studios from 1936 to 1965. *Academy of Management Journal*, **39**(3), 519–43.

Mudambi, R. (2008). Location, control, and innovation in knowledge-intensive industries. *Journal of Economic Geography*, **8**, 699–725.

Oakley, K. and O'Connor, J. (2015). *The Routledge companion to the cultural industries*. London: Routledge, Taylor & Francis Group.

Peteraf, M.A. (1993). The cornerstones of competitive advantage: A resource-based view. *Strategic Management Journal*, **14**(3), 179–91.

Potts, J. and Shehadeh, T. (2014). Compensating differentials in creative industries and occupations: Some evidence from HILDA. In G. Hearn, R. Bridgstock, B. Goldsmith and J. Rodgers (eds), *Creative work beyond the creative industries: Innovation, employment and education* (pp. 47–60). Cheltenham, UK and Northampton, MA, USA: Edward Elgar Publishing.

QUT (n.d.-a). Defining the creative economy. Retrieved from https://research.qut.edu.au/creativehotspots/defining-the-creative-economy/.

QUT (n.d.-b). The creative economy in Australia. Retrieved from https://research.qut.edu.au/creativehotspots/wp-content/uploads/sites/258/2019/10/Factsheet-1-Creative-Employment-overview-V5.pdf.

QUT (n.d.-c). The creative economy in Australia: Cultural production, creative services and income. Retrieved from https://research.qut.edu.au/creativehotspots/wp-content/uploads/sites/258/2019/10/Factsheet-2-Employment-by-sector-V5.pdf.

Scott, A. (2004). Cultural-products industries and urban economic development: Prospects for growth and market contestation in global context. *Urban Affairs Review*, **39**(4), 461–90.

Solow, R. (1987). We'd better watch out. *New York Times Book Review*, 12 July, p. 36.

Wilterdink, N. and Form, W. (2018). Social structure. *Encyclopædia Britannica*, 26 April. Retrieved from https://www.britannica.com/topic/social-structure. (Last accessed 27 November 2019.)

Workplace Gender Equality Agency (2019). *Gender pay gap*. Retrieved from https://www.wgea.gov.au/terms/gender-pay-gap. (Last accessed 17 January 2020.)

3. The relationship between creative employment and local economies outside capital cities[1]

Greg Hearn,[2] Stuart Cunningham,[3] Marion McCutcheon[4] and Mark David Ryan[5]

INTRODUCTION

A substantial amount of debate about the creative economy has centred on the network effects of the location of talent in large cities, implying that, creative agglomeration is an urban phenomenon of first-tier cities (e.g., Hall, 1998). Cultural amenity in inner-city areas, from museums and theatres to galleries and live performance venues, have also been argued to be a primary driver of location decisions for the *creative class* (e.g., Florida, 2002). Moretti's (2012) study *The New Geography of Jobs* suggests that there is a relationship between the role of creative workers in new knowledge industries and employment growth in the United States (US). Moretti's data shows that US employment growth is largely restricted to brain-hub cities – cities with the highest concentrations of highly specialised innovation workers that generate more local jobs for service workers – and these have a higher density of creative class workers. Haskell and Westlake (2018) suggest that growth in intangible investments by companies has led to a growing inequality between urban and other regions.

This focus on cities as the centres of the creative economy has, however, been challenged by research (e.g., Faggian et al., 2013; Flew, 2012; Gibson, 2012; Waitt and Gibson, 2009) focusing on the reasons why creative workers choose to live and work in small urban centres, regional cities and the suburbs. In addition, there is a growing body of literature on the negative effects of inner-urbanist creative economy policies (Oakley and Ward, 2018). Different outcomes for creative graduates in different regions have been analysed by Faggian et al. (2013). Certainly, there are many creative workers outside major cities. In any jurisdiction, the ratio of city creatives to those outside cities depends of course on total population outside cities. This in turn depends on the aggregation of industries and their labour demands outside cities. In

Australia, this varies significantly state by state. In Greater Sydney, the heart of Australia's creative economy, there are around 125 000 workers employed in creative occupations, and 16 000, 11 per cent of the state's creative workforce, elsewhere in New South Wales (QUT Digital Media Research Centre, n.d.-a). In Queensland, 32 per cent of the state's 53 000 creative workers live outside the Greater Brisbane area.

What does this mean for the future of creative work? Like cities all over the world, Australian cities struggle with too much population growth, rising housing costs and the changing nature of employment. Urban population growth also puts pressures on infrastructure, environment, and social and community resilience. The liveability of cities is a growing global problem. At the same time, regional and rural communities fight to maintain a sustainable population and to retain people with experience and skills. Flagging economic growth during a protracted and countrywide drought is pushing governments to look for job creation opportunities outside traditional economic sectors such as agriculture. Given that, in most democratic jurisdictions, non-urban electorate populations are smaller to offset the challenges faced in servicing more dispersed citizens, there remains a bottom-line political imperative to continue to support non-urban areas and invest in their capacity to create jobs. Many non-urban local governments around the world now incorporate creative industries policies in their economic development strategy. However, very little evidence exists on creative industries intensity and its relationship to local regional economic factors. In this chapter, we present preliminary findings of an ongoing study into the dynamics of the creative economy outside the major capital cities of Australia.

To understand whether there is a future for creative work outside major cities, we focus on the relationship between creative employment and Gross Regional Product (GRP)[6] outside the capital cities of Australia. There are two broad hypotheses that have informed our thinking that higher levels of creative industry activity might be associated with higher levels of GRP. First, it stands to reason that the greater the GRP, the more jobs there are for creative workers. Certainly, this is true in cities. The reasons for this are complex, but two of many possible drivers could be more discretionary income for consumption of creative products, and a higher business need for creative services such as advertising, design and digital software. The second hypothesis is that creative work in a region may contribute to the GRP growth in a region. This could be achieved, for example, by accelerating the market share of business through advertising (e.g., Bayazit and Genc, 2019) or by increasing a region's innovation capabilities to create *new* products or businesses (e.g., Lobo et al., 2014). If these hypotheses are both true, over time we would expect to see an increasing returns dynamic (Arthur, 1996); that is, increases in GRP support the generation of creative jobs, which create more GRP, which creates more jobs, and so on. This is exactly what Østbye et al. (2018) investigated in analysing population and regional employment data of 250 regions in Finland, Sweden and Norway. The strongest statistical effect was that people go to regions where new jobs have been created (i.e., where GRP growth is strong). That is, people move into regions in order to get a job. However, Østbye et al. also found a secondary effect of relevance here. First, they found that creative class[7] jobs are created when workers in general move into a region. They also found that growth in creative class jobs and other occupations are bi-directionally correlated. That is, creative class jobs growth is followed by growth in other occupational jobs, but also vice versa.

The ambition and available data for this chapter are more modest and limited than Østbye et al. (2018) had, but speak to similar conceptual issues. Our aims are first to investigate whether the positive correlation between GRP and creative employment holds for non-urban areas of Australia. Put simply: Are there more creatives employed in high GRP regional economies? Secondly, we ask how creative workers might actually help GRP to grow and thus create jobs for others, which then creates more jobs for creatives in a reciprocal cycle of expansion.

To pursue this analytical agenda, we utilise Australian national quantitative employment and GRP measures.[8] The quantitative creative employment data central to this chapter is derived from the 2016 Australian Census. The national census, officially titled The Census of Population and Housing, administered by the Australian Bureau of Statistics (ABS), is conducted every five years, tracks employment by industry and occupation, and, as such, provides an opportunity to investigate creative-workforce dynamics outside of capital cities. Here, we compile employment data for creative occupations coded using the Australia and New Zealand Standard Classification of Occupations (ANZSCO) (see Hearn and McCutcheon, 2020) in all local government areas (LGAs). It is important to note that this employment data includes both creative occupations in specialist creative organisations as well as creative occupations *embedded* in other sectors of the economy (see Hearn and McCutcheon, 2020). We sourced National Institute of Economic and Industry Research estimates of GRP for LGAs from the regional economic and demographic consultants, .id consulting (2019). While examining trends in one LGA can identify trends in employment and GRP, examining the relationship between the two across a national data set of employment measures and GRP measures for all LGAs allows us to reveal differences in trends across different region types. In parallel with the quantitative work reported here, the authors are conducting qualitative studies of the creative economy in varied LGAs (from peri-urban areas, second-tier cities, towns and remote areas) across the five mainland states of Australia. In this chapter, we also use three diverse regions in the state of Queensland (Central West, Cairns, Sunshine Coast) to exemplify results of the quantitative analysis and gain insight into the possible underlying mechanisms by which creative workers might add value to business.

CREATIVE EMPLOYMENT AND GRP IN NON-METROPOLITAN REGIONS

The Creative Intensity of the Workforce Measure

In addition to gross employment counts, we also developed a measure of the creative intensity of the workforce, the relative number of creative workers as a ratio of the total workforce, using the same ANZSCO data as the employment counts. Creative intensity can be thought of as a quantitative snapshot of the creative DNA of creative work in a locale, pinpointing numerically where the comparative strengths and weaknesses in a local creative economy may be found. For illustrative purposes, Table 3.1 provides a comparison of creative intensities for the three Queensland regions, and inner Sydney as a benchmark for creative intensity. Unsurprisingly, inner Sydney, the heart of Australia's creative economy, evidences higher creative intensity in all categories. For example, the creative intensity of *software and digital content* is nine times that of the Sunshine Coast, which is in turn twice that of Cairns. *Architecture and design* and *advertising and marketing* are also significantly higher in inner Sydney than in the Sunshine Coast and Cairns. In inner Sydney, software and digital content has the highest creative services intensity, followed by advertising and marketing, with architecture and design third. In Cairns and the Sunshine Coast, architecture and design is predominant, with advertising and marketing second and software and digital content third. In the Sunshine Coast, in contrast to the Central West and Cairns, software and digital content intensity has grown strongly, more than tripling between 2011 and 2016. According to the census statistics, there were no full-time creative services workers in Central West Queensland in 2016.

The cultural production creative intensities in inner Sydney are less than half those for creative services. Nevertheless, they are among the very highest in Australia, up to around four times higher than the three case sites, with the exception of *visual arts*, which is similar across all four sites. Notably, although having no creative services workers, the Central West is relatively strong in all cultural production intensities. Noting that absolute numbers are very small, nevertheless, creative intensities for *film, TV and radio*, for *music and performing arts* and for *visual arts* all grew in the Central West between 2011 and 2016, despite severe economic contraction.

Table 3.1   *Regional creative intensities compared with Sydney, 2011*
                and 2016 (%)

	Central West		Cairns		Sunshine Coast		Inner Sydney	
Occupation sector	2011	2016	2011	2016	2011	2016	2011	2016
Creative services								
Advertising and marketing	0.11	0.00	0.35	0.35	0.45	0.53	2.27	2.67
Architecture and design	0.00	0.00	0.58	0.56	0.81	0.86	2.23	2.27
Software and digital content	0.00	0.00	0.14	0.19	0.14	0.44	3.13	3.98
Cultural production								
Film, TV and radio	0.19	0.33	0.15	0.13	0.16	0.15	0.99	0.88
Music and performing arts	0.08	0.24	0.22	0.24	0.22	0.26	0.61	0.54
Publishing	0.48	0.36	0.26	0.21	0.38	0.31	1.36	1.05
Visual arts	0.00	0.12	0.11	0.08	0.21	0.18	0.13	0.10
Creative occupations	0.99	0.90	1.82	1.76	2.66	2.73	10.72	11.48

Note: All figures are percentages of total employment, regions are defined by LGA boundaries.
Central West Queensland includes Blackall–Tambo Regional Council, Longreach Regional
Council and Winton Shire Council. Sunshine Coast includes Sunshine Coast Council and Noosa
Shire Council. Inner Sydney refers to the City of Sydney LGA, which comprises only the central
business district (CBD) and some surrounding suburbs.
Source: ABS (2016). Findings based on use of ABS TableBuilder data.

Correlations Between Employment Measures and GRP Measures

A correlational analysis establishes whether there is a significant directional
relationship between two variables. It does not prove there is a causal relationship, of
course. However, where there is no significant correlation, the possibility of a simple
direct causal relationship can be eliminated. Correlational analysis is therefore
a useful preliminary step in understanding, formulating and excluding hypotheses
to be tested with more advanced statistical methods such as multiple regression and
cross-lagged longitudinal analyses. Since national correlational analyses of creative
industries employment and GRP have never been conducted before in Australia, we
approached this exercise as a preliminary exploration for the purpose of advanc-
ing our understanding of the two driving aims of the investigation.

Using data for all 486 LGAs in Australia, for the census years of 2011 and
2016, we established correlations between 2016 creative employment counts and
creative intensities, and GRP (2016) and GRP growth (2011–2016). We grouped
creative occupations into the following categories using ANZSCO (see Hearn
and McCutcheon, 2020; QUT Digital Media Research Centre, n.d.-b):

- advertising and marketing;
- architecture and design;

- software and digital content;
- film, TV and radio;
- music and performing arts;
- publishing;
- visual arts.

Elsewhere, we have developed the analytical distinction between predominantly business-to-business creative services occupations (advertising and marketing; architecture and design; and software and digital content) and cultural production occupations predominantly involved in final consumption, or business-to-consumer services (film, TV and radio; music and performing arts; publishing; and visual arts) (Cunningham, 2014). This distinction is useful for examining how the creative industries connect with the greater economy, where they might contribute to other industries, and how they drive consumer demand. We also segmented the dataset of 486 LGAs based on their relative remoteness: very remote (74), remote (71), outer regional (159), inner regional (94) and city (88).[9]

As a baseline, we examined relationships between creative occupation employment (counts and creative intensities) and GRP for all LGAs in Australia (including cities). Examining first the relationship between creative employment counts and GRP, Table 3.2 shows that all categories of creative occupation employment counts correlate positively with GRP. This is unsurprising because LGAs with higher GRP would be expected to have higher employment across the economy. Correlations between the number of creative occupation employment counts and compound average annual GRP growth between 2011 and 2016, however, show a divergence among different categories of creative occupation:

- Weak positive correlations between GRP growth and employment counts in the occupation categories:
    - advertising and marketing;
    - architecture and design;
    - film, TV and radio;
    - music and performing arts;
    - publishing;
    - visual arts.
- No relationship between GRP growth and the software and digital content category.

The correlation between GRP measures and creative intensities (i.e., the number of creative occupations as a proportion of total employment) reveals different relationships from that between GRP measures and employment counts. The relationships between creative occupation intensity and total GRP show:

- Strong positive correlation between GRP and creative intensity for software and digital content occupations.
- Medium positive relationship between GRP and the following categories:
  - architecture and design;
  - advertising and marketing;
  - publishing.
- Weak positive correlation between GRP and the following categories:
  - film, TV and radio;
  - music and performing arts.
- No relationship between GRP and visual arts intensity.

Considering the relationships between creative intensity and GRP growth shows:

- Weak positive correlations between GRP growth and occupation intensity in the following creative categories:
  - advertising and marketing;
  - software and digital content;
  - architecture and design;
  - publishing.
- No relationship between GRP growth and occupation intensity in the following categories:
  - film, TV and radio;
  - music and performing arts;
  - visual arts.

Next, we turn to correlation calculations segmented in terms of LGA size and region type, as described above, focusing on creative intensity measures and GRP measures. In contrast to the national data set analyses above, there were no significant correlations between creative intensities and GRP growth in any of the region types. There were, however, significant correlations between creative intensities and GRP per se, and they suggest a different set of dynamics in different regional types (see Table 3.3). In summary:

- In very remote areas, there does not appear to be any relationship between creative intensity and GRP, reflecting the high relative variation in regions with very low employment counts.
- Creative intensity of the software and digital content category has the strongest correlation with GRP across all regional segments except outer

regional, where advertising and marketing, and music and performing arts, are similar, if not a little higher.

- Correlations with GRP are usually higher for creative intensities of creative services occupations than cultural production occupations in all regional segments, but there are exceptions; for example, publishing is higher than architecture and design in some segments.
- The highest correlation between the creative intensity of the film, TV and radio category and GRP is in remote areas.
- The only significant correlation between GRP and the creative intensity of the music and performing arts category is in outer regional areas.

Table 3.2   *Correlations between creative employment counts and intensities, and GRP and GRP growth, for all Australian LGAs*

Occupations	Counts and GRP	Counts and GRP Growth	Intensity and GRP	Intensity and GRP Growth
Creative services				
Advertising and marketing	++++	+	++	+
Architecture and design	++++	+	++	+
Software and digital content	++++	0	+++	+
Cultural production				
Film, TV and radio	++++	+	+	0
Music and performing arts	++++	+	+	0
Publishing	++++	+	++	+
Visual arts	++++	+	0	0

Notes:
All correlations are positive and are significant at least at the $p < .05$ level.
++++ = Very strong: $r > .75$
+++ = Strong: $r > .5$
++ = Medium: $r > .25$
+ = Weak: $r < .25$
0 = No correlation.
Source: GRP data from .id consulting (2019). Employment data from ABS (2016) using ABS Table Builder data.

Table 3.3   *Correlations between GRP (total) and creative intensities for different regions, 2016*

Occupations	Major cities	Inner regional areas	Outer regional areas	Remote areas	Very remote areas
Creative services					
Advertising and marketing	++	++	++	++	0
Architecture and design	0	0	+	++	0
Software and digital content	+++	++++	++	+++	0
Cultural production					
Film, TV and radio	0	+	0	++	0
Music and performing arts	0	0	++	0	0
Publishing	++	++	0	0	0
Visual arts	0	0	0	0	0

Notes:
All correlations are positive and are significant at least at the $p < .05$ level.
++++ = Very strong: $r > .75$
+++ = Strong: $r > .5$
++ = Medium: $r > .25$
+ = Weak: $r < .25$
0 = No correlation.
Source: GRP data from .id consulting (2019). Employment data from ABS (2016) using ABS TableBuilder data.

CASE STUDIES OF REGIONAL LGAS

To further explore explanations for these dynamics, we now turn to qualitative considerations of three diverse regions in Queensland: the Central West, Cairns, and the Sunshine Coast. These three sites are different in terms of their remoteness from the state's capital city (Brisbane), their size, their economic composition, and the state of the economy in terms of employment, creative intensities and GRP. One feature they had in common, as revealed in the qualitative research, was the importance of creative intensity in tourism, although their tourism assets and the markets they serve are different in each case. This diversity proved important in understanding how creativity intensity has multiple mechanisms that contribute to GRP. In each case, site interviews were conducted in 2019 using a convenience sample of selected key informants (e.g., local government representatives, industry organisations, creative workers, venue representatives, innovation brokers and managers) (Cunningham et al., 2019a; 2019b; 2019c). The interviews canvassed a range of creative work issues, but in relation to this investigation, we looked for examples of how creative workers were either creating new businesses, or helping existing businesses to increase their incomes and, thus, creating more employment.

Central West: A Distressed Agricultural Economy that Imports its Creative Workforce to Bolster the Economy

The Central West region is an outback community, 700 km from the coast and 1000 km from Brisbane. It comprises the towns Longreach, Winton and Blackall-Tambo, which have been in drought for the last five years. The population of the Central West is 13 318 and full-time total employment is 3352. With agriculture at the core of its economy, between 2016–17 and 2017–18, GRP in Longreach declined by 18.2 per cent and in Winton by 12.0 per cent, while employment in the region fell by an average of 2.0 per cent per year between the national census years of 2011 and 2016. In Longreach, agriculture continues to be the largest employer, accounting for 20.1 per cent of local jobs in 2016. However, 9.6 per cent of jobs in Longreach are estimated to be in tourism-related roles, making tourism the town's fifth-largest employer (REMPLAN, 2018). In terms of full-time creative employment, the number of creative workers in the Central West is very small: only 30 workers, all in cultural production occupations. In 2016, there were no full-time creative services workers in the Central West because there simply is not the volume of stable work to warrant more than freelance engagement (Cunningham et al., 2019b). As measured by our approach in this chapter, in 2016, the Central West had a total creative intensity of 0.90 per cent. And yet, this case can tell us much about the role of creative workers in remote communities, the contribution of creative activity to GRP, the creation of employment, and how creative employment in the future may be created in such remote communities. The reason for this is that despite the low numbers of local creative workers, the Central West has a very significant economic footprint in museum tourism, film-location provision, and film-related tourism, art and music festivals.

The local museums provide cultural experiences through local stories and are crucial tourism drivers, in addition to eco and outback tourism. Although reliant on federal, state and local government grants and donations to cover investments in infrastructure, the museums are sustainable, with day-to-day running expenses covered by ticket sales, restaurants and cafes, and gift shops.

The museums provide full-time and ongoing employment, and skill development opportunities for volunteers. Local museums include:

- The Australian Stockman's Hall of Fame in Longreach;
- The Qantas Founders Museum;
- Winton's Waltzing Matilda Centre.

The region is also an established, and an increasingly popular, location for film and television production; and the economic flow-on benefits, though sporadic, are important injections into the local economy. Although there is no significant film or television production industry in the region, Winton has provided the outback landscape for domestic and international screen productions since 2005, for example, the United Kingdom and Australian co-production *The Proposition* (2005), the Australian film production *Goldstone* (2016) and the six-part Australian Broadcasting Corporation (ABC) drama television series *Black B*tch* (2019). The US mini-series *Texas Rising: The Lost Soldier* was filmed in the region for the History Channel in 2015. These productions result in economic flow-on effects in the form of accommodation for cast and crew, catering, employment for locals as extras, and carpentry, among many other examples. Winton's Vision Splendid Film Festival showcases outback-themed movies from around the world, attracts national and international film production students from numerous universities annually, offers short film courses, and aims to generate ongoing economic benefits for Winton.

There are important implications for the aims of this chapter. First, this shows how the creative intensity measure based on full-time local creative employment has potential to belie the significance of creative work and its contribution to local economies. It is clear from this case that part-time and fly-in/fly-out creative work also needs to be considered. This is also one possible explanation for the lack of correlation between GRP and creative employment in remote communities in the quantitative models in this chapter. Further, even in this remote distressed community, there is evidence that creative enterprises and activities contribute directly to local total employment, and contribute to GRP through their own revenues; they increase demand and revenues for businesses in other sectors such as accommodation and retail; and they provide income to their own employees. There is anecdotal evidence that outback tourism is increasing demand for digital marketing, web development and online booking systems (Cunningham et al., 2019b). A small number of advertising and marketing professionals are embedded in other industries in the region (Cunningham et al., 2019b).

Cairns: A Stable Diverse Economy with High Creative Intensity in Tourism

Cairns is situated on the coast, between the UNESCO-listed World Heritage Daintree Rainforest and Great Barrier Reef, and is the second-largest city in Northern Australia. Cairns is a highly multicultural city, with high proportions of Indigenous Australian people and people from neighbouring Papua New Guinea and other Melanesian countries. With a steady population and falls in employment, the Cairns economy nevertheless grew by an annual average of 1.8 per cent between the census years of 2011 and 2016 (ABS, 2016; .id consulting, 2019). The Cairns international airport allows the city to function as a central hub for Far North Queensland and a popular destination for conferences and sporting events (Cairns Regional Council, 2017a; 2017b). Cairns has a very diverse economy with strengths in health and social services, tourism, construction, agriculture and marine industries (including shipbuilding), but education also makes an important contribution. The marine sector is large and includes a major fishing fleet, reef tourism, military marine infrastructure, and high-end super yachts. That is, there is economic activity across the value chain in Cairns for the marine industry.

With a population of 156 900, and total employment of 67 460 in 2016, the city's 1190 creative workers give Cairns a total creative intensity of 1.8 per cent – double that of the Central West. There are 738 creative services workers and more than half of them work in industries other than the creative industries. Architecture and design has the highest creative intensity of all creative occupations in Cairns, and is above average compared to other Australian non-urban centres. Architecture and design workers are mainly employed in specialist firms rather than embedded in other sectors. Conversely, a large majority of advertising and marketing workers are embedded in other sectors rather than in specialist firms. The proportion of people working in software and digital content occupations in Cairns (0.19 per cent) is a little lower than the average for all LGAs outside Brisbane (0.26 per cent) (QUT Digital Media Research Centre, n.d.-a). Cairns has a relatively high number of cultural production workers, with music and performing arts and publishing having the highest creative intensity. Forty-five per cent of cultural production workers are employed in sectors other than the creative industries, which is high compared to the rest of Australia.

Tourism is an important exemplar of creative intensity in Cairns. The most recent industry statistics show that tourism-spending in Cairns in 2017 exceeded AU$2.1 billion – nearly two-thirds of tourism expenditure for all of Far North Queensland – with 2.8 million visitors contributing 8.2 per cent of its economic value added, and tourism-related industries employing 9000 people. There are significant numbers of advertising and marketing workers and cultural production workers in the sector. However, after 30 years of growth, infrastructure bottlenecks, limited numbers of flights into Cairns, withdrawal of services of major carrier Cathay Pacific in 2019, natural disasters and outdated marketing strategies have seen Cairns' share of the tourism market fall over the last 10 years (Cunningham et al., 2019a). There is also widespread concern about the future of the Great Barrier Reef. Cultural tourism experiences are becoming more significant in this context.

Tjapukai Aboriginal Cultural Park has grown over three decades, and is an outstanding example of Indigenous tourism. Owned by Indigenous Business Australia, the majority of staff at Tjapukai are Australian Indigenous people, with most identifying as Djabugay. The enterprise's cultural content, messaging and Dreamtime stories are the intellectual property of the Djabugay people from the rainforest of tropical North Queensland and are tightly controlled under Indigenous protocol. Other Indigenous entities add to the cultural tourism offering, including:

- UMI Arts;
- Gimuy Fish Festival;
- AppOriginee: A new initiative by local Traditional Owner Gudju Gudju and partners.

Eco-tourism has been very important to Cairns and directly employs creatives, particularly in traditional advertising and marketing, but Cairns is facing challenges that require it to diversify. Without this, the revenues of local tourism firms and their contribution to GRP could fall. This could be achieved, for example, by adding new experiential products in eco-tourism, and expanding offerings in Indigenous cultural heritage, agri/food tourism, and creative digital and entertainment experiences. Tourism drives accommodation, which drives some of the demand for architecture and design, particularly for large mixed developments. But the cultural tourism strategy needs to differentiate itself internationally through high-end design values and local authenticity in accommodation amenity – that is, a distinctive tropical architectural and design style, informed partly by Indigenous traditions. This also helps to grow the already strong local design sector. In addition, the building of impressive cultural venues has also contributed to this trend, as well as providing significant general employment locally. Supporting this main dynamic is the significance of design in marine industries. Creative services employment right across the diverse economy is evident, particularly in advertising and marketing.

Sunshine Coast: A Booming Diverse Economy with an Innovation Agenda

The Sunshine Coast[10] is a family-friendly and lifestyle-oriented locale, popular both as a tourist and residential destination. The Sunshine Coast has one of the fastest-growing economies in Australia. Its population is growing strongly and local government is working to establish world-class communications, transport and health infrastructure, while maintaining the integrity of the region's highly valued environment and lifestyle. The local government has an explicit creative industries development strategy, in recognition of the council's role in attracting and retaining the skilled and talented individuals who it anticipates will be important for the future of the economy. Health, social services, construction and tourism-related sectors are the largest sectors of the Sunshine Coast economy, with secondary strengths in education and professional services.

The Sunshine Coast has a population of 346 512 and employment of 130 978. There are 3575 creative workers, with a creative intensity of 2.7 per cent, which is one-and-a-half times that of Cairns. The rank order of creative intensities for advertising and marketing and for architecture and design is the same as Cairns. Similarly, architecture and design employment is primarily in specialist firms, and the advertising and marketing workforce is embedded in other sectors. This is probably due to the same tourism-driven dynamic in Cairns. However, the software and digital content creative intensity is twice that of Cairns and tripled between 2011 and 2016. The creative intensities in the cultural production sector are very similar to Cairns, with many creatives working in other sectors. There are strengths in music, performing arts and visual arts.

There is a wider range of entrepreneurial creative industry businesses in the Sunshine Coast region than in the other two cases. This is some evidence that lifestyle is an attractor for such firms to relocate to the area. Talent attraction is an explicit strategy of the local government. This is part of the innovation and entrepreneurship ecosystem strategy facilitated by local government and industry. Established in 2017, the #SCRIPT (Sunshine Coast Regional Innovation Project Team) programme focuses on smart cities, food and agri-business, health and wellbeing, sustainability and environment, and creative industries (Cunningham et al., 2019c). The activation of the strategy has included the establishment of networks, brokerage, investment, mentoring, incubation, talent attraction programmes and spaces. The Refinery, a creative incubator, is a related business development programme aimed at assisting arts and cultural practitioners to build sustainable businesses, including in the fashion, retail, e-commerce, education, environment and First Peoples sectors.

In addition to specialist cultural production firms, there is evidence of firms seeking to provide *enabling inputs* into other sectors. Near Field Creative is a key example of the enabling input model. The company started by creating a clothing label targeted at gamers in order to establish a manufacturing supply chain, before embedding wearable technology that was useable by other small business, sporting clubs and not-for-profits. The wearable technology enables point-of-customer-contact monetary transactions and data compilation. There are several companies that are attracting national and international attention for innovative products that rely on design inputs, for example, Helimods (helicopters), Praesidium Global (military unmanned ground vehicles) and Smartline Machinery (smart cabinets for the health sector).

As in the other case sites, tourism is one of the most significant industries in the region. Attracting over three million visitors each year, it is the third-largest industry in the Sunshine Coast LGA (.id consulting, 2019; Yigitcanlar et al., 2018). Its family-friendly reputation, excellent beaches and relaxed lifestyle make it an ideal holiday destination for Australians. However, it is reasonable to suggest that the character of tourism in the Sunshine Coast is different from that of Cairns. Cairns and the Sunshine Coast have approximately the same number of visitor nights (10 million in 2018/19). However, Cairns is significantly more international in its tourism market (45 per cent of visitor nights) compared to the Sunshine Coast (12 per cent). While the Sunshine Coast has a very strong domestic market for tourism (both intra- and inter-state tourism), it lacks the world-leading eco-tourism attractions of Cairns. The relevance of arts and culture to tourism also differs between the two cities. For example, Indigenous cultural offerings on the Sunshine Coast, while very important to local cultural identity, are in a nascent stage in terms of scale compared to Cairns. International tourists are more likely to seek out cultural experiences than domestic tourists (Tourism Research Australia, 2018). This different tourism dynamic may mean different pathways to economic impact are operating in the two cities. Specifically, compared to Cairns, whose population is essentially static, the population of the Sunshine Coast is growing strongly. It is reasonable to suggest therefore that advertising and marketing of the Sunshine Coast destination serves a dual purpose, promoting tourist visits as well as attracting residents.

We found evidence for this in our interviews with the Sunshine Coast media sector. The Sunshine Coast is an important hub for regional Queensland broadcasting services: the ABC, Grant Broadcasters and EON Broadcasters operate radio studios, Network Seven operates a television studio in Maroochydore, and community broadcasters are based in Buderim. This provides employment for film, television and radio professionals that is above average outside metropolitan centres. These media channels often feature lifestyle content that indirectly promotes the Sunshine Coast as a destination, enhancing popular perceptions of the area's natural environment, community belonging and cultural identity. It is reasonable to suggest this boosts both tourism and population growth.

The significance of creative intensity to GRP in the Sunshine Coast arguably replicates some trends evident in Cairns. Creative services intensity is found across many non-creative sectors and therefore is growing the revenues of businesses and creating more jobs, which produce income inputs to GRP. Cultural production creative intensity is similar and contributes to vibrancy and activation of entertainment and retail offerings and spaces. The media sector promotes the region as a destination. However, there are some findings suggestive of additional elements. Population growth is an important driver of GRP in the Sunshine Coast. There is stronger creative intensity in architecture and design, which is arguably driven by the domestic construction sector. This is a case in which GRP growth leads to the need for more creatives. There is a more explicit emphasis on innovation in the Sunshine Coast, which is important to GRP in that it may create new businesses for a region, and we found evidence that this is beginning to occur. There is a strong and growing software and digital content ecosystem, which provides services to a range of industries, as well as starting new businesses. Local procurement of these services means more employment and income, contributing to GRP. Finally, creative talent attraction is an explicit strategy of local government, with the belief that this will be good for both innovation and creative industries growth. Cultural and lifestyle amenity, which creative workers contribute to, is important for talent attraction.

DISCUSSION AND CONCLUSIONS

It is important at the outset to reiterate the caveats about what analyses the data and methodology do and do not allow. The quantitative data are snapshots at particular points in time (i.e., 2011 and 2016) and correlational analysis does not allow any causation (except null causal relationships) to be inferred. In relation to the GRP growth variable, in particular, inferences about its causation are meaningless anyway, since it essentially precedes employment measures. However, it is still possible to treat this as a lead indicator for creative intensity. The same sequence issue applies to qualitative data because it was collected in 2019. Nevertheless, the data does allow consideration of meaningful questions such as:

- whether larger economies employ more creatives;
- what the intensity of creative work is outside capital cities;
- whether GRP growth is a lead indicator of creative intensity;
- whether causal links between creative intensity and GRP measures for particular regions or occupations can be excluded or deemed unlikely.

Similarly, because the qualitative interviews addressed retrospective and prospective matters in addition to the current "state of play", such interviews do provide exploratory consideration for the purpose of understanding and exemplification of possible causal dynamics. In terms of causal dynamics, it is important to note that revenues of businesses and regional personal incomes are important inputs into GRP; therefore, mechanisms by which businesses grow revenue and create jobs are very relevant to our considerations. With these constraints and opportunities for analysis in mind, we now turn to a holistic exploration of the research agenda.

Analysis of the unsegmented national data set showed there is a very strong positive correlation between GRP and total creative occupation employment for *all categories* of creative work. This means that strong GRP is associated with high levels of creative employment. Because the data set is based on a national census, our findings delineate that the creative economy is not limited to, and operates outside of, major capital cities. Put simply, creative workers benefit from strong economies. This is still the case even if the most likely direction of causality in this correlation is only from GRP to employment. Prosperous economies have more money to buy art and cultural products, and lead to more businesses that need creative services; therefore, strong economies are good for creative employment.

In terms of GRP *growth*, the relationship to creative employment is more complex in the national data set, but we still found weaker though positive correlations with most creative services and cultural production occupations. This again is good news overall for creatives. The weakness of the positive correlations, when compared with the correlations with GRP per se, is another important fact here. If GRP growth is perfectly correlated with GRP size, this result would simply be an artefact of size. However, in the data set, there is only a weak correlation between GRP size and growth. This means that high growth or low growth does not always relate to high or low creative employment in every LGA. Indeed, as will be discussed, when regional segmentation is considered, a different set of relationships between economic factors and creative employment emerges. Another notable and surprising finding is that there is no relationship between GRP growth and software and digital content employment counts. Later discussion suggests why this might be.

When we consider the national data on creative intensity (i.e., the relative numbers of creative workers to total employment) and GRP, we again find positive correlations between GRP and all categories of creative employment (except visual arts). This means that larger economies not only have more creatives, but also have relatively more of them than other occupations. *This effect is stronger for creative services occupations, and is most strong for software and digital content*. The effect is similar for relationships between creative intensities and GRP growth, though the strength of the relationships is weaker. More insight into the relationship between creative services intensities and GRP is provided in the regional segmentation analyses and case studies.

The advertising and marketing category operates at the consumer capture end of supply chains to drive demand and support revenue generation, and the category's relationship with GRP measures is strong both nationally and in all regional sectors, excluding very remote areas. There is reasonably robust, well established and evenly distributed advertising activity at the regional and local level, but not as population levels decline in very remote areas. In the case studies, we found evidence for this in both Cairns and the Sunshine Coast, where there are high levels of embedded advertising and marketing professionals across sectors. Tourism, which is a business-to-consumer sector, is an important part of the local economy in both these regions and depends heavily on advertising to promote both the region as a destination as well as specific attractions. Even in the Central West, where census data shows no creative services workers, our interviews located a small contingent of part-time and full-time advertising and marketing workers.

The architecture and design category operates at the start of supply chains and may add value by expanding product offerings through differentiation. Many designers, however, are employed in somewhat replicative activities, for example, in standard domestic and commercial architecture. This may explain why, in general, this category's relationship to GRP measures is not as strong as other creative services. Creative intensities only correlate with GRP in outer regional and remote areas and this may simply be a case of local procurement being more convenient than outsourcing, which is common in larger regional centres. In the case studies, we found strong specialist architecture and design sectors in both Cairns and the Sunshine Coast. Our interviews in Cairns suggested that distinctive local tropical design was an important consideration in the development of both publicly funded cultural venues as well as signature accommodation buildings. We found Indigenous interior designers who were seeking to bring cultural motifs into their work. Cairns has an explicit strategy to be distinctive in promoting itself to the international tourism market, which accounts for 45 per cent of visitor nights (.id consulting, 2020). In the Sunshine Coast, we suggested that population growth and domestic architecture were a driver of a demand for architects and found less evidence for an emphasis on the construction of signature buildings and a distinctive regional design style. There were however examples of new product development through industrial design in manufacturing, defence and health.

The software and digital content category potentially affects all parts of a supply chain, offering general purpose technologies that increase value differentiation, reduce production and increasingly reduce distribution costs, and increase market channels. This category mostly has the strongest creative services correlation with GRP measures nationally and in all regions, excluding very remote areas. Notably, however, gross employment numbers in the software and digital content category do not correlate with GRP growth in the national data. This may mean that in aggregate, software and digital content may increase productivity through driving efficiencies and reducing employment, but may also increase employment as digital products and consumer channels expand. For example, routine digital systems such as inventory and financial management require fewer digital workers once installed and have a longer life span. Creative digital products and market channels require constant refreshing to maintain consumer interest.

In the three case sites, we found quite different digital ecosystems. In the Central West, interviews suggested there was a digital skills shortage and that this was a frustration for many local businesses. The growing outback tourism sector has a demand for digital skills in digital advertising and creative digital tourism experiences, but meets this through contracts with suppliers out of the region. Cairns has relatively low intensity in this category, and some interviewees suggested this was a barrier to value-adding to tourism, for example, through digital memorabilia and virtual experiences. The Sunshine Coast has high intensity in this category and an explicit development strategy for its digital ecosystem, and we found examples of both efficiency enhancing services, as well as innovative development of new products and services, for example, in wearable technology and entertainment.

There are also significant differences in the relationship between cultural production employment and GRP across the regional segments in our quantitative analysis. For example, the highest correlation between the creative intensity of the film, TV and radio category and GRP is in remote areas. The Central West is suggestive of why this may be the case. Potential locations for film and television production are a natural asset in many remote regions, and the town of Winton has capitalised on this in a significant way and has built a sustainable film festival on the back of these locations. These activities make a valuable contribution to the sustainability of the Winton economy and employment. Furthermore, the only significant correlation between GRP and the creative intensity of the music and performing arts is in outer regional areas. The attendance of live music and performance arts depends on local consumption dynamics and factors such as discretionary income, job security and confidence in the economy – factors that are more robust in larger local economies. Access to affordable housing, studio space and lifestyles may also be implicated. The relationship between creative intensity and GRP measures for publishing is stronger nationally than all other cultural production categories. However, the explanation for this is different from all other creative occupational groups.

Publishing comprises journalists, authors, library workers and curators. It is the only creative occupational category that declined between 2011 and

2016 nationally. In the regional segments the positive relationship to GRP only applies in major cities and inner regional areas. This is possibly a story of contraction and centralisation of a workforce. For example, as regional media outlets such as the ABC and commercial providers have been closed, there have been relatively fewer job losses in the major urban and larger regional towns and cities. Overall, we can say that for almost all major socio-economic indicators, regional, rural and remote populations lag major urban populations. There are also dramatic correlations between degree of remoteness and cost of cultural consumption in a country where population is radically thinly dispersed outside major urban centres (Cunningham and Potts, 2011). The variations in correlations between cultural production creative intensities and GRP in different regional segments also may be explained by these factors.

In conclusion, our data and analysis suggest that the creative economy is not merely an urban phenomenon. We have shown that, in Australia, there is strong empirical and qualitative evidence that the creative economy can be found in LGAs from second-tier cities to remote communities, and that creative workers make valuable economic contributions when economies are distressed as well as when they are booming. This is important to the consideration of the future of creative work. Although the urban concentration of creative work is likely to accelerate, we found evidence that creative intensity is important to non-urban areas as well. Furthermore, creative economies have different dynamics in regions with different size populations, GRP and distances from major cities. In general, though, creative employment correlates very strongly with GRP for all categories of creative occupation and there will be future opportunities for creative work in many different community contexts. For creative services, we identified possible mechanisms to explain this, namely, a possible bi-directional relationship between creative services growth and GRP growth, though it is more likely that business growth is the more powerful element. That is, GRP growth is a lead indicator of creative services work. Our evidence also suggests that as local creative economies grow in scale, the creative DNA of the creative economy evolves. This can be tracked by the creative intensity of each creative occupation – that is, the percentage of creative workers compared to total employment. This new measure also indicates possible strengths and weaknesses of the local creative economy. Software and digital content is a notable example in our three case sites; the Sunshine Coast has rapidly grown its intensity, while Central West and Cairns find lack of digital skills an issue. This measure can distinguish between creative employment in specialist firms, and creative work embedded in other sectors of the economy. In the case sites, we found evidence of strong embedded workforces in advertising and marketing, most likely predominantly in tourism. For cultural production work, the future is likely to depend more on the particular industry sector and the specific region. Cultural production work in regional and remote communities has some advantages in sustainability, due to the role it plays in maintaining a cohesive community and local identity. As evidenced by the three case sites, in Australia, tourism, and in particular cultural tourism, may be an important growth sector for the future creative workforce outside capital cities, with many cultural production workers finding employment in this sector. While urbanisation will continue, a key question for the future of creative work outside mega-cities is how small cities with relatively low creative intensities can grow to a greater size where there is a higher concentration of work for creatives. Our chapter has developed important insights into how this evolution could happen.

NOTES

1. This chapter was completed with funding from the Australian Research Council Linkage project (LP160101724) led by Queensland University of Technology in partnership with the University of Newcastle, Arts Queensland, Create NSW, Creative Victoria, Arts South Australia and the WA Department of Culture and the Arts. It uses excerpts from Cunningham et al. (2019a; 2019b; 2019c) completed for that grant. The views expressed are those of the authors alone.
2. ORCID iD: 0000-0003-2245-3433. Creative Industries Faculty, Queensland University of Technology, g.hearn@qut.edu.au.
3. ORCID iD: 0000-0002-7437-1424.
4. ORCID iD: 0000-0003-3755-9070.
5. ORCID iD: 0000-0002-1544-1007.
6. It is important to be clear that GRP is affected by many factors exogenous to any region in question, for example, international competition, drought, energy costs and so on.
7. Their classification of jobs aligns with Florida (2002) and includes jobs focused on here, but a broader set of knowledge workers.
8. Australia is a modern economy, which, though significantly dependent on resources for its export performance, has a diverse range of other sectors and services. In addition, Australia has a well-established public and private sector in arts, culture, media and creative industries more broadly. It also has a number of contrasting city-level economies, ranging from Sydney, which is a typical global urban creative ecosystem, through to a wide range of regional second-tier cities and remote Indigenous communities that reflect a condition similar to developing countries in some cases. The economy of Australia is in the top 20 in terms of its size. *And as such, the analysis represents a proxy for a wide range of contexts that are relevant globally.*
9. Region types are defined here according to the Australian Statistical Geography Standard (ASGS) Remoteness Structure, which allocates SA1 areas to a remoteness category based on its road distance to the nearest urban centre. We allocated each LGA to a remoteness category according to the category of the majority of the SA1 areas that fell within its boundaries. Source: QUT Digital Media Research Centre (n.d.-a).
10. In this study, the Sunshine Coast includes the two LGAs of Sunshine Coast and Noosa.

REFERENCES

Arthur, W.B. (1996). Increasing returns and the new world of business. *Harvard Business Review*, **74**(4), 100–109.

Australian Bureau of Statistics (ABS) (2016). *Census of population and housing.* Retrieved from TableBuilder: https://www.abs.gov.au/websitedbs/censushome.nsf/home/tablebuilder.

Bayazit, Z. and Genc, E.G. (2019). An analysis of reciprocal influence between advertising expenditures and Gross Domestic Product. *International Journal of Economics and Financial Issues*, **9**(2), 41–7.

Cairns Regional Council (2017a). *2016–2017 Annual Report.* Retrieved from https://www.cairns.qld.gov.au/__data/assets/pdf_file/0019/221743/annrep1617.pdf.

Cairns Regional Council (2017b). *Corporate Plan 2017–2022.* Retrieved from https://www.cairns.qld.gov.au/__data/assets/pdf_file/0004/209722/CorpPlan17_22.pdf.

Cunningham, S. (2014). Creative labour and its discontents: A reappraisal. In G. Hearn, J. Rodgers, B. Goldsmith and R. Bridgstock (eds), *Creative work beyond the creative industries: Innovation, employment and education* (pp. 25–46). Cheltenham, UK and Northampton, MA, USA: Edward Elgar Publishing.

Cunningham, S. and Potts, J. (2011). The price of our great digital divide. *The Australian*, 14 May, p. 4.

Cunningham, S., McCutcheon, M., Hearn, G., Ryan, M.D. and Collis, C. (2019a). Australian cultural and creative activity: A population and hotspot analysis: Cairns. Retrieved from QUT Digital Media Research Centre website: https://research.qut.edu.au/creativehotspots/wp-content/uploads/sites/258/2019/12/Creative-Hotspots-CAIRNS-report-FINAL-V1-20191220.pdf.

Cunningham, S., McCutcheon, M., Hearn, G., Ryan, M.D. and Collis, C. (2019b). Australian cultural and creative activity: A population and hotspot analysis: Central West Queensland: Blackall-Tambo, Longreach and Winton. Retrieved from QUT Digital Media Research Centre website: https://research.qut.edu.au/creativehotspots/wp-content/uploads/sites/258/2019/11/Creative-Hotspots-Central-West-Queensland-FINAL-20190801-web.pdf.

Cunningham, S., McCutcheon, M., Hearn, G., Ryan, M.D. and Collis, C. (2019c). Australian cultural and creative activity: A population and hotspot analysis: Sunshine Coast. Retrieved from QUT Digital Media Research Centre website: https://research.qut.edu.au/creativehotspots/wp-content/uploads/sites/258/2019/11/Report-Sunshine-Coast-FINAL-20191021-Reduced-web.pdf.

Faggian, A., Comunian, R., Jewell, S. and Kelly, U. (2013). Bohemian graduates in the UK: Disciplines and location determinants of creative careers. *Regional Studies*, **47**(2), 183–200.

Flew, T. (2012). Creative suburbia: Rethinking urban cultural policy – the Australian case. *International Journal of Cultural Studies*, **15**(3), 231–46.

Florida, R. (2002). *The rise of the creative class: And how it's transforming work, leisure, community and everyday life.* New York, NY: Basic Books.

Gibson, C. (ed.) (2012). *Creativity in peripheral places: Redefining the creative industries.* London, UK and New York, NY, USA: Routledge.

Hall, P. (1998). *Cities in civilisation: Culture, innovation, and urban order.* London: Weidenfeld and Nicolson.

Haskell, J. and Westlake, S. (2018). *Capitalism without capital: The rise of the intangible economy.* Princeton, NJ, USA and Oxford, UK: Princeton University Press.

Hearn, G. and McCutcheon, M. (2020). The creative economy: the rise and risks of intangible capital and the future of creative work. In G. Hearn (Ed.), *The future of creative work: Creativity and digital disruption*. Cheltenham, UK and Northampton, MA, USA: Edward Elgar Publishing, pp. 14–33.

.id consulting (2019). Australian economic indicators. Retrieved from https://content.id .com.au/economic-indicators-australia.

.id consulting (2020). Cairns Regional Council: Economic profile. Retrieved from https://economy.id.com.au/cairns/tourism-visitor-summary. (Last accessed 20 February 2020.)

Lobo, J., Mellander, C., Stolarick, K. and Strumsky, D. (2014). The inventive, the educated and the creative: How do they affect metropolitan productivity? *Industry and Innovation*, **21**(2), 155–77.

Moretti, E. (2012). *The new geography of jobs*. New York, NY: Houghton, Mifflin, Harcourt.

Oakley, K. and Ward, J. (2018). Creative economy, critical perspectives. *Cultural Trends* [Special Issue], **27**(5), 311–12.

Østbye, S., Moilanen, M., Tervo, H. and Westerlund, O. (2018). The creative class: Do jobs follow people or do people follow jobs? *Regional Studies*, **52**(6), 745–55.

QUT Digital Media Research Centre (n.d.-a). Data tables. Retrieved from https:// research.qut.edu.au/creativehotspots/data-tables/.

QUT Digital Media Research Centre (n.d.-b). Defining the creative economy. Retrieved from https://research.qut.edu.au/creativehotspots/defining-the-creative-economy/.

REMPLAN (2018). Economy, jobs and business insights. Retrieved from: https://www .economyprofile.com.au/longreach/tourism/employment.

Tourism Research Australia (2018). Local Government Area profiles, 2017: Cairns (R), Queensland. Retrieved from https://www.tra.gov.au/ArticleDocuments/312/Cairns %20(R).xlsx.aspx.

Waitt, G. and Gibson, C. (2009). Creative small cities: Rethinking the creative economy in place. *Urban Studies*, **46**(5–6), 1223–46.

Yigitcanlar, T., Kamruzzaman, M., Buys, L. and Perveen, S. (2018). *Smart cities of the Sunshine State: Status of Queensland's local government areas*. Retrieved from https://eprints.qut.edu.au/118349/.

4. A taxonomic structural change perspective on the economic impact of robots and artificial intelligence on creative work[1]

Ben Vermeulen,[2] Andreas Pyka[3] and Pier Paolo Saviotti[4]

INTRODUCTION

Recent publications on the job-destroying potential of robots and advanced software such as artificial intelligence (here referred to collectively as R&AI) have revived classical inquiries into whether technological change leads to the *end of work* or, rather, whether a *rebound* of, or a *structurally lower* rate of, employment is to be expected. Whenever robots become more nimble, dexterous and easily configurable, AI becomes more sophisticated, and R&AI technology becomes better adjusted to the (capricious) human environment, more and more tasks currently performed by humans become susceptible to substitution (cf. Brynjolfsson and McAfee, 2011; Ford, 2015; Frey and Osborne, 2017). For instance, robots were previously mostly applied in manufacturing sectors, but recent technological advances have increased applicability to the extent that robots are now also introduced in agriculture, logistics, maintenance, health care and other sectors. Moreover, the span of sectors in and breadth of tasks to which R&AI technologies are applied may well increase over the forthcoming years.

That said, there are numerous countervailing forces that have, historically, restored employment, increased wages and improved working conditions (Acemoglu and Restrepo, 2016; 2017; 2018; Vivarelli, 2007; 2013). For instance, whenever automation makes products or services cheaper, demand increases, which effectively increases demand for labour. Additionally, automation may increase productivity and make complementary (possibly new) skills more valuable (e.g., configuring robots, training AI), which may lead to increases in wages and thereby consumption, and may thus create employment.

On top of that, in capitalist economies, competition of firms drives erosion of profit margins, which triggers firms to look for new sources of income by creating new products, services and business models (Schumpeter, 1942). This may give rise to breakthrough technologies and new sectors 'mopping up' the newly unemployed (Pasinetti, 1981; Saviotti and Pyka, 2004; 2008; Vermeulen et al., 2018). Given that these new sectors initially host labour-intensive work, employment is effectively reinstated (cf. Acemoglu and Restrepo, 2018).

However, technological change affects not only the employment rate, but also the task content of jobs. Over the last couple of centuries, many countries have seen a transformation from a mostly agricultural economy to a manufacturing economy and, now, to a service economy (cf. Fisher, 1939). The end-of-work perspective argues that R&AI may gradually take over physical work in agriculture and manufacturing *and* knowledge-based work in services, such that there are no tasks left for humans to fulfil.

Under the assumption that R&AI is adopted purely to increase productivity (thus, ignoring other reasons such as window-dressing and techno-chauvinism), cost-economic rationales can be used to tell which tasks will be left to humans, which skills are thus in demand, and thereby how wages are affected. The main conjecture of this chapter is that, with increasing adoption of R&AI technologies, there is an increasing prominence of tasks in which humans excel (e.g., creativity), in which technological solutions are excessively expensive (e.g., physical 'super'-dexterity, operating in a highly complex, unpredictable environment), or in which humans reject technological solutions or prefer a human touch (e.g., requiring social skills).

This chapter focuses on the economic impact of R&AI *on creative work specifically*, pertaining to work taking place in separate creative sectors, but, importantly, also (complementary) tasks performed in other sectors. To this end, we propose a taxonomy of sectors based on the relationship with the technology in question. That is, whether sectors make or apply R&AI, whether services or products from these sectors are supporting or competing with R&AI, or whether the sectors merely receive income from all these other sectors. The various mechanisms changing the employment rate and tasks executed are positioned within this taxonomy.

In particular, this chapter proposes a framework of mechanisms driving gradual structural change of an economy in which creativity plays an increasingly prominent role. Notably, there are three drivers of structural change, which, in conjunction, determine whether economies develop into the end-of-work, rebound or structurally lower scenario: *adoption of R&AI* causes shifts in labour demand, *innovation* creates new employment opportunities, and *education* facilitates labour mobility. Subsequently, the role of countries' institutional frameworks is highlighted, particularly with regard to the effect of labour-market flexibility, education and innovativeness in structural change.

In addition, the chapter stresses that technological change affects the competitive position of countries, both due to the institutional framework and the prior labour-market composition. As such, for instance, countries in the global production chain offering low- to middle-skilled, routinised production work may be hard hit and lose their competitiveness.

Conclusively, under the conjectured increasing prominence of complementarities such as creative work (and social skills, physical super-dexterity and so on), both in *applying* and *emerging* sectors, any inherent limitations to educationability (of creativity and other complementarities), or institutional impediments to labour mobility, technological change and innovativeness, may jeopardise economic growth, and cause unemployment and inequality.

TAXONOMIC PERSPECTIVE ON STRUCTURAL CHANGE

In many countries the employment rate, hours worked and task content of jobs have changed substantially over the last couple of centuries. The Fisher–Clark–Fourastié model describes how economies developed from being dominated by activities in agriculture, later in manufacturing and finally in services (see Fisher, 1939). This phenomenon of *structural change* in the composition of the labour force across sectors is long studied (*nota bene* Baumol, 1967; Kuznets, 1966; Pasinetti, 1981). Scholars in the Schumpeterian perspective attribute structural change to technological change (see Silva and Teixeira, 2008). More specifically, structural change is brought about by competition as the subsequent erosion of profit margins forces firms to introduce labour-saving, productivity-enhancing technology. However, ultimately, firms are also forced to develop and make new products and services that cater to the needs of (potential) customers; in turn, this increases the demand for labour. Without the creation of new products and services, and the resulting customer demand and ultimately labour required, the productivity-enhancing technology would cause mass unemployment (which would require unemployment benefits and income redistribution; see, for example, Leontief, 1983). However, over the past centuries, rather than seeing mass unemployment caused by technological revolutions, economies have adapted (albeit sometimes after social reforms) and new employment opportunities have emerged (Mokyr et al., 2015), with notably an absolute increase in demand for high-skilled, knowledge-intensive and non-routinised work.

Arguably, in the Schumpeterian perspective, the application of R&AI technologies may be a prominent driver for contemporary structural change. The current vintages of R&AI technologies are primarily production equipment that increase labour productivity of, and reduce demand for, labour-intensive, routinised work in predictable environments (cf. Autor et al., 2003). As such, there may be an *absolute* decrease in demand for such routinised work, or rather, work in general, and a *relative* increase in demand for more knowledge- or intellect-intensive, non-routinised work and low-skilled work that is hard to automate (see Goos et al., 2014). However, there are some concerns that the impact of R&AI on employment is different from that of previous waves of technological change. In particular, regarding structural change, it remains to be seen whether the emergence of new sectors is as massive as with the radical up-scoping of products manufactured and services provided in the past. Here, a close-up is taken of the employment effects of the introduction of R&AI technologies and how the labour-saving effects are offset by increases in demand for labour.

Taxonomy of Sectors and Countervailing Forces

The introduction of R&AI will affect tasks performed, skills required, jobs in demand and, thereby, (inequality in) wages earned, (un)employment rates, education required and job opportunities for people. A prominent contention about the impact of R&AI implementation is that, although the increase in pro- ductivity does reduce the demand for (particularly types of) labour, this may be compensated by the creation of new tasks performed, skills required and jobs in demand (cf. Acemoglu and Restrepo, 2016; 2017; 2018; Vivarelli, 2007; 2013). In particular, the introduction of (new) technology increases demand for complementary tasks, skills and jobs (Goldin and Katz, 1998; Griliches, 1969). Moreover, new technology may decrease unit costs: firms may lower their product prices, and as a result, bring about an increase in product demand, which may increase labour demand. The new production technology may also just replace existing technology, and thus may merely *deepen automation* rather than replace labour (Acemoglu and Restrepo, 2018). In addition, notably high-wage jobs in high-technology or knowledge-intensive sectors are accom- panied by *spillover demand* in other sectors (cf. Goos et al., 2018; Moretti, 2010), such that there are *local multipliers* in employment and wages.

We argue that the impact of R&AI differs across sectors, and we propose to turn matters inside-out by discerning different types of sectors by looking at their relationship with R&AI technology. More specifically, we formulate a taxonomy based on the relationship with the focal technologies and thereby distinguish the *making, applying, supporting, competing* and *spillover* sectors. The taxonomy of sectors may be used as a focal lens to clarify how R&AI is substituting labour, and how this is compensated and complemented (cf. Vermeulen et al., 2018). With an increase of implementation in the applying sectors, there is a definite increase in labour demand in the making sectors engaged in researching, developing, building, implementing and servicing R&AI. This may well be at the expense of demand for labour in the competing sectors, such as those producing machines, tools or equipment; however, there may also be a (temporary) increase in investments in these sectors to stay ahead, catch up or leapfrog R&AI technologies (see Ward, 1967, for this so-called *sailing ship effect*).

In the applying sectors, in which firms acquire and apply R&AI, progressive rationalisation of manufacturing processes may indeed save labour. However, it is often specific tasks that are completely displaced, not workers. After all, with the introduction of new technology, there may well be new and complementary tasks for workers to perform (e.g., controlling, configuring, maintaining robots, training and validating AI). These tasks may even give rise to new occupations. Indeed, in particular sectors applying R&AI, such as health care, education and hospitality, there may primarily be a shift in the task sets rather than net loss of jobs, with people able to focus on 'what really matters', such as face-to-face interaction with patients, knowledge transfer to students, welcoming guests and so on. This is also discussed in more detail in the context of what we call the *escalation of dimensionality* introduced later. Moreover, the loss of jobs due to rationalisation may well be partially offset by a decrease in the unit cost of products, which leads to an increase in product demand, and subsequently, demand for labour.

In sectors supporting the applying and making sectors – for example, by means of training and education, consulting and legal support – there may be a (possibly temporary) surge in labour demand to facilitate the technological transition, and certainly a partial structural shift in subjects addressed. Finally, the spillover sectors provide products and services that are not *directly* affected by the introduction of R&AI (e.g., tasks might not be suitable for robotisation). However, the employment and wages in spillover sectors are affected by the consumption of workers employed in the other types of sectors and, as such, by the total employment and disposable income of the workers in these sectors.

In fact, innovation of various sorts may ultimately culminate in new industrial and service activities so distinct that they emerge as new sectors that create work for (some of) the unemployed.

Impact on the Creative Sectors and Creative Tasks in Other Sectors

Triggered by the adoption of R&AI and the substitution of labour, the various complementary and countervailing forces also contribute to the structural change across the various sectors. Here, we use the taxonomic perspective to study the effects on creative work of introducing R&AI through shifts in

productivity, product prices, employment, wages, and task and skill requirements. We discern four key effects:

1. There is substitution of tools and equipment in the creative sectors as the
   applying sectors, which may change (a) skills required, introducing new
   ones, transforming or shifting prominence of existing ones, and (b) productivity and possibly the products and services provided and subsequent
   labour demand. Notably, the actual creative aspects, rather than mere
   transformation of material (or data), may become more prominent.
2. The application of R&AI in non-creative sectors changes income and
   employment, and therefore, demand for goods and services provided by the
   creative sectors. Moreover, there is a shift in tasks performed by humans,
   with likely increasing prominence of human creativity.
3. The increasing scale and scope of application increases demand for, and
   wages of, labour in research, development and production of R&AI in the
   making sectors, which in turn affects demand for other goods and, in turn,
   labour. Moreover, some effects are predominantly related to the transition
   from old to new technologies, such as a shift in and temporary surge in
   demand for education and consulting, and investments in the competing
   sectors to keep up.
4. The creative sectors are affected by changes in spillover demand from
   changes in total income in other sectors. Moreover, labour-market competition may flare up in the applying or in the creative spillover sectors
   depending on whether R&AI induces demand for creatives or rather creates
   unemployment among people seeking to move into creative sectors.

Each effect is summarised in Table 4.1 and discussed in more detail in the
following sections.

Application in creative sectors

While it is hard to delimit the set of creative sectors, we refer here, broadly,
to traditional media (film, TV, video, radio, photography), new media and
entertainment (computer games, rendered movies, blogging/vlogging), crafts,
fashion, toys and games, industrial design, architecture, advertising and
marketing, print and digital publishing, cultural sectors (museums, galleries,
libraries), gastronomy, and a few others. The (potential) impact of the introduction of R&AI technologies in the creative sectors (i.e., regarding these as
applying sectors) may well be quite diverse.

First, while there may be direct applications of R&AI technologies in
the creative sectors, currently their impact is limited. Because there are few
relevant contemporary examples, we have to resort to examples from the introduction of computer technology, in the 1980s, for instance. Arguably, while

Table 4.1   *Summary of effects of engagement with R&AI in the various types of sectors on the demand for creative work (and thereby, indirectly, wages)*

Sector	Change in employment (for workers with creative skills) and reasons
Making	+ (Potential) diversification of applications and deepening automation in applying sectors → research & development of new R&AI technology (e.g., health-care service robots, AI evaluation and testing, data analysts)
Applying: creative	+/− Change in tools, devices and transformation of tasks, not pure substitution of work (e.g., physical editing of film to using software)
	+ Culmination in new products (e.g., in the past: 3D artists, game designers)
	? Culmination in new media and business models may disrupt industry (e.g., entrants replace incumbents?); net effect unclear
	+/− Local multiplier: total disposable income in other sectors affects demand for products of the applying and (non-)creative sectors
Applying: creative and non-creative	− Substitution of tasks
	+ Emergence of creative tasks and embedded creatives as complementary to job; doing what human is better at than R&AI
	+/− Increasing productivity → decreasing unit cost → increasing product demand → increasing labour demand
Supporting	+/− Change in topics to teach and consult; provide advice on both in the applying and making sectors; at least temporary surge facilitating the technological transition
Competing	− R&AI substitute
	+ Investments to stay ahead, catch up or leapfrog R&AI
Spillover: creative and non-creative	+ Local multiplier: total disposable income increases in the making sectors
	+/− Local multiplier: total disposable income changes in the applying, supporting and competing sectors
	− Unemployment in the applying sectors may increase competition for (particular) jobs in creative sectors

Notes: A '+' ('−') means that the demand for creative work is increasing (decreasing) because of the reason stated, while '+/−' indicates that there are opposite effects at work. The table includes concepts from Vermeulen et al. (2018).

computerisation has affected the creative sectors, we have found no evidence (primary data or research reports) of massive outright substitutions. R&AI has instead transformed occupations by, for example, moving to a new medium or providing new tools for the same task. For instance, editing movies transformed from cutting and pasting film at an editing table to using software on digital clips. Similarly, design of toys, architectural models and so on is done in 3D (three-dimensional) software and using 3D printing, rather than making physical prototypes. Writing a book transformed from going to the library to do research and typing out handwritten notes on a typewriter, to searching online and using word-processing software.

Secondly, on top of transformations of tasks, computerisation calls for complementary skills (e.g., computer-aided design requires design and computer skills) and created many new occupations (e.g., 2D and 3D computer graphics artists, 3D modellers, computer music artists, web designers, bloggers/vloggers). This of course brings about a shift in skills to be taught and consulting to be provided. There are also new occupations emerging, which may even co-exist with 'traditional' occupations. Moreover, the new technology may shift attention from 'artisanal' or 'handicraft' activities to more creative activities, and the net effect is not clear a priori. For instance, algorithms do contribute to media editing (see, for example, Napoli, 2014), but the impact on actual employment is unclear.

Application in non-creative sectors

Despite the broad range of R&AI applications in creative sectors, the impact seems to be limited due to the nature of the tasks. While future technological developments may well change this, at present R&AI is mostly applied in non-creative sectors that allow for rationalisation and progressive optimisation of routinised, repetitive tasks. However, the application of R&AI need not necessarily destroy jobs outright, but may well bring about a transformation to the task set, notably to ensure proper implementation and exploitation of R&AI technology. Indeed, one of the most prominent countervailing forces mentioned in literature is the increased demand for workers with complementary skills. R&AI arguably is incapable of certain social interactions meaningful to people and may fall short on certain forms of creativity, physical super-dexterity and intellectual flexibility. As such, particular tasks, and therefore certain jobs requiring these skills, continue to be done by humans. Production and service processes may be increasingly organised around these 'human' skills. Consequently, creativity may emerge as a particularly valued complementary skill, and creative occupations may well be *one of the (last) bastions of employment* in the applying sectors, even if economies face the end of work.

Making and supporting sectors

Any (prospective) increase in the *application* of R&AI creates jobs in robotics (e.g., mechatronics engineering), AI software engineering, and researching and developing advanced new applications. Arguably, new jobs in the making sectors contribute to employment beyond these jobs themselves. First, there are local multiplier effects causing an increase in labour demand; each newly created job in a region may create additional jobs due to the increase in demand for tradable goods. The local multiplier effects may contribute as many as five additional (lower skilled) jobs for each high-skilled/high-tech job (Moretti, 2010), although this number varies with regional circumstances (Goos et al., 2018). A critical note, though, is that jobs in R&AI may themselves be substitutes for jobs in machine-building and software-engineering industries. Secondly, the transition from using old technology to new technology in the applying sectors will affect activities such as education, consulting and legal advice, and may thus induce a temporary surge in employment in the supporting sectors.

Local multiplier effects on the creative sectors
Whenever the total income in a region changes (due to changes in employment and wages) in the making, applying and supporting sectors, there may be different effects on employment and wages in creative sectors.

First, due to the aforementioned local multiplier effect, changes in employment, wages, spare time and disposable income of workers in these making, applying and supporting sectors change demand for products and services from, and therefore employment and wages of, workers also in the creative sectors. In fact, goods and services from the creative sectors may have high-income elasticity, and moreover – due to non-homotheticity – may only be in high demand whenever incomes are sufficiently high. Consumption of products and services from the creative sectors may be localised and thus may rise and fall strongly with total income in the region.

Secondly, there are effects of changes in competition for jobs. Whenever robotisation does indeed induce unemployment, workers trained/educated and (previously) employed in the applying sectors may well vie for jobs in the creative sectors. Generally, under sufficient labour mobility, competition for particular (lower skilled) jobs in the creative sectors may become fierce, driving wages down. Alternatively, if demand for creative workers increases in the applying sectors, the scarcity in creative sectors may increase, driving wages up.

New sectors
On top of the developments within and across *existing* sectors, there is emergence of *new* sectors, many of which cannot even be properly named or circumscribed yet. Innovation of various sorts may ultimately culminate in new industrial and service activities so distinct that they emerge as new sectors that create work for (some of) the unemployed. What these new sectors provide in terms of products and services may often be hard to envision in advance. For example, in the 1950s, few people predicted that information and communication technology would give rise to the many sectors it has, would grow to the size it has, with applications in the many and widespread ways it has. These new products, new services and new sectors are a fundamental source

of economic growth (Pasinetti, 1981; Schumpeter, 1942). Initially, these new sectors often boast labour-intensive work (cf. Acemoglu and Restrepo, 2018). This *reinstatement effect* may well be required for a balanced growth path (Acemoglu and Restrepo, 2018). Do note, however, that there may be a considerable disparity between the skills of the unemployed and the skills required for emerging occupations. Indeed, those becoming unemployed may do routinised tasks in predictable environments, while emerging occupations may consist of high-skilled, knowledge-intensive tasks. That said, perhaps not all vacancies can be (immediately) fulfilled, and neither will the (newly) unemployed from the applying sectors immediately find a job without up- and re-skilling. As firms are unlikely to hire such unemployed people for the emerging occupations, the public sector may have to provide social benefits or educational facilities for up-skilling (see also the next section on this).

Quite a few of the sectors that have emerged in the last couple of decades actually started out as niches in the creative sectors (e.g., computer games, digital graphics, computer music, rendered movies) but have matured, and some have even outgrown the sector of origin. As such, the creative sector is a prominent (and possibly increasing) source of product and service innovation.

Additional Causes for the Rise of Prominence of Creative Work

As argued, creative work may well be one of several pivotal complementarities emerging upon progressive rationalisation of production and services. Moreover, demand for products from the creative sectors depends, to a certain extent, on total disposable income in the other sectors. However, there are additional potential causes for increasing prominence of creative work: (a) progressive institutionalisation and intensification of research and development activities; and (b) escalation in the product dimensions at stake in competition.

First, in the Schumpeterian perspective, R&AI technologies are not only introduced as a labour-saving, productivity-enhancing and cost-cutting measure. These technologies and other technologies with R&AI at their core (e.g., autonomous vehicles, warehouse logistics, surveillance systems, tailored education, and potentially many others yet unconceived) give rise to new products and services, business models and possibly whole new sectors. Research and development of these products has been and is further institutionalised. Indeed, labour statistics show an increase in employment in scientific, technical and research services, and therefore, arguably, work revolving around human ingenuity and creativity.

Secondly, there are two prominent by-products of increasing global competition that put more emphasis on creativity and thereby on employment in creative occupations *across all sectors*. The first by-product is that there is a process of *escalation of dimensionality* of the features of products and services. We argue that, in a capitalist economy, there is a relentless drive to shake off competitors, which causes an endogenous re-setting of the pivot in competition. As such, firms evolved from focusing almost exclusively on price in the 1960s, to including product quality in the 1970s, to including time-to-market and lead time (requiring production flexibility) in the 1980s, culminating in customisability and newness (requiring innovativeness) in the 1990s. Today, there is an increasing prominence of design, marketing, appeal, aesthetics, image, symbolic knowledge and generally 'newness' included in the dimensions that firms compete on. As such, the importance of and employment in the final stages of the value chain have already increased and are expected to increase further. The second by-product is that incessant product and service innovation has become a key competitive strategy; therefore, firms require dynamic innovation capabilities. As a result, firms are likely to grow and nurture capabilities to think creatively and come up with new product and service concepts, and new business models and ways of engaging customers. This will, obviously, increase the value of creative skills and abilities, and increase the demand for creative people in every sector. Therefore, *embedded creatives* are increasingly prominent (Cunningham and Higgs, 2009; Hearn and Bridgstock, 2014).

FUTURE SCENARIOS

As argued in the previous section, the adoption of R&AI contributes to the increasing prominence of creative work across the various types of sectors. In fact, creative work (together with a narrow range of other complementary tasks) may well be one of the last bastions of employment. This section contains three sub-sections. The 'Scenarios' sub-section focuses on the *total rate of employment* and provides three baseline scenarios on employment, with explicit regard for creative work. The 'Institutional and Structural Mediators' sub-section discusses how institutions in market economies moderate the impact of technological change on structural change, and how varieties of market economies may hence be affected differently. Moreover, it is discussed how technological change may thus also indirectly alter international relationships of countries. The 'Limits to Change?' sub-section discusses the primary variables driving developments in the direction of either one of the future scenarios, thereby explicitly regarding the market economies' institutions and structuralist interactions between countries.

Scenarios

As mentioned before, when it comes to employment rate, there are two extreme scenarios: the end of work, in which there is *mass unemployment,* and the rebound, in which the labour market returns to *near-full employment* after an initial dip. There is one other scenario, the middle ground, the structurally lower scenario (see Vermeulen et al., 2018). Each scenario is now discussed with particular focus on the rate of employment in creative work.

The first scenario is that of the end of work, in which R&AI technology is replacing human labour at such a rate that the additional work due to counter-vailing effects and new tasks, occupations and sectors does not offset the loss of employment. This net destruction of jobs may have several causes (possibly in conjunction). First, the new creative work initially induced by the intro-duction of R&AI is ultimately *also* substituted by new vintages of R&AI. So, over time, these technologies have to take over not only the complementary tasks in the applying sectors, but also work in the supporting and competing, and even the making sectors (i.e., each thus becomes an applying sector as well). Moreover, for this scenario to materialize, R&AI should also replace the complementary tasks that are newly emerging in the applying sectors (e.g., face-to-face interaction, knowledge transfer, hospitality, creativity, etc.). So, the scope of tasks taken over by R&AI technology has to increase, particularly by increasing dexterity, flexibility and social comprehension. Robots become so much more nimble, flexible and easily configurable, and AI becomes so much more sophisticated, that virtually all tasks currently performed by humans become susceptible to substitution (cf. Brynjolfsson and McAfee, 2011; Ford, 2015; Frey and Osborne, 2017). Secondly, other sources of new jobs (notably innovations) generate too few jobs or the opportunities cannot be reaped. The latter may be the case if education and training changes too slowly and labour mobility is hampered. Indeed, the employment rate reflects the 'race' between the job destruction by process rationalisation and automation on one hand, and the job creation and opportunity reaping potential by educa-tion on the other (Acemoglu and Restrepo, 2016; Brynjolfsson and McAfee, 2011; Goldin and Katz, 2007; Keynes, 1930; MacCrory et al., 2014). In this line of thought, R&AI technologies ultimately can and do handle all the work (cost economically), including social work, health care and psycho-social support, and with regard to the (narrowly defined) creative sectors also creat-ing art, composing and making music, writing books and plays, and directing and 'generating' movies.

The second scenario is that of the rebound, in which, as in the end-of-work scenario, the employment rate may experience a decline due to rapid substi-tution, but now ultimately the rate at which new tasks, jobs and sectors are created and exploited gives rise to increasing employment. Essentially, this net job creation has two requirements. The scope and scale of cost-economically viable and technically feasible application of R&AI does, at some point in time, increase more slowly than the creation of these tasks, jobs and sectors. Clearly, this may be caused by high specificity or complexity of the R&AI required for these tasks. This, in turn, requires that both the creation of these new tasks, jobs and sectors, and the supply of labour to fulfil demand for these jobs, are not stifled. To this end, education is to ensure sufficiently high mobility of workers, and consulting, incubation and funding are to enable creation and exploitation of new technological opportunities. Note that while several countervailing forces increase labour demand (Acemoglu and Restrepo, 2016; 2017; Vivarelli, 2007, 2013) – including the *growth bonus* of spillovers from the making sectors multiplying labour demand in other sectors (cf. Fagerberg, 2000; Goos et al., 2018) – the creation and reaping of new opportunities is crucial to restore labour demand (cf. Acemoglu and Restrepo, 2018). So, despite the increase in productivity, high rates of and structural change in employment may be more due to shifts in demand for goods and services. This, in turn, is to be attributed to non-homothetic preferences – that is, changes in demand with changing incomes. In this line of reasoning, local creative and cultural sectors may thus benefit from rising productivity when coupled with rising incomes.

The third scenario is that in which the employment rate levels off at a structurally lower level. Over the last few centuries, the employment rate has bounced back after each technological revolution. However, this was accompanied by a structural change in work performed by human workers. Moreover, generally, the hours worked per annum have also decreased substantially (while the disposable income has increased). As argued before, with the introduction of R&AI, particular types of work may be left to humans, i.e. creative work, care and service work in which humans want to interact with other humans, and work requiring extreme physical dexterity or intellectual flexibility that is yet too costly to develop R&AI for. As such, there is a subset of work that will be left to humans. It is hard to predict, though, how many humans would be employed and how many hours they would work. In one possible extreme case, well-paid, full-time jobs may be available only to a small number of people and the rest would be unemployed or have precarious employment. Another potentially more likely scenario is that the available work hours may be spread across the population in a more or less equally distributed way, as is currently the case. However, those becoming unemployed may well need to be (re)educated and (re)skilled for jobs in demand. Note that with increasing disposable income, humans may just be satisfied with the work performed and consumption opportunities available, and therefore, may prefer to work less than looking for even higher incomes, thereby effectively moderating the emergence of new sectors and technological progress.

It is important to note that regardless of which one of the latter two scenarios on the employment *rate* becomes reality, there may also be polarisation in the *composition* of the labour market. Because the routinised tasks in predictable environments in particular are automated, and apparently these are found in the middle-skill segments, application of R&AI is expected to unleash fierce competition for physical, low-skilled work and, to a lesser extent, for knowledge-based, high-skilled work (Autor et al., 2006; Goos et al., 2009; 2014).

Institutional and Structural Mediators

The impact of R&AI on employment in general, and the prominence of creative work in particular, is mediated both by (a) the institutional framework in place and (b) the shift in competitive position of products and the change in the composition of the labour force of countries in structuralist, world-systemic interactions.

The first key factor is that the institutional framework of a country mediates the impact of R&AI on the rate and direction of structural change. For example, think of how labour unions, labour-market regulations, social-security arrangements, the organisation of education and the coordination of innovation affect the emergence of new sectors, labour mobility, wage inequality and therefore consumption opportunities. The varieties of capitalism perspective (Hall and Soskice, 2001) classifies market economies by the institutional framework for coordination of the various actors facing, among others, structural and technological change. The two main forms of market economies, the liberal and coordinated market economies, differ notably on the regulation of the labour market (e.g., in terms of the degree of unionisation, employment protection), social security (e.g., benefits, assistance in education and finding jobs) and meso-level organisation of innovation activities (e.g., competition versus collaboration). Labour-market regulation constrains the substitution of labour with R&AI. Moreover, labour-market regulation in conjunction with institutional provisions for education also affects the labour mobility possible, the abilities to exploit technological opportunities and the responsiveness to fluctuation in demand.

Proponents argue that labour-market deregulation and flexibility increase the propensity to hire people upon (foreseen) demand swings in existing sectors and for more risky endeavours in emerging sectors (under uncertain future demand). As such, the supposed propensity to hire people stimulates the take-off of new sectors and, as a result, structural change, whereby employment is more likely to 'bounce back'. *Opponents* argue that deregulation and flexibility increase the likelihood that firms will substitute workers who are already disadvantaged, that is, those lower skilled, in lowly paid routine work.

These may have substantial difficulty in finding work in occupations that are in demand. Therefore, labour-market flexibility may induce unemployment, a widening skill-gap between the employed and unemployed, and increasing inequality in incomes and wages due to which further demand is also negatively affected (see, for example, Dosi et al., 2018). It is stressed here that institutional arrangements and policy measures to retain such routinised work may (a) harm international competitiveness of output and (b) hamper market forces moving labour in the direction of new, emerging opportunities and thereby jeopardise long-term economic growth and increase the risk of falling behind.

While it may well be true that newly unemployed people lack skills for work in new sectors, and thus form a source of inequality, there are policy measures and institutional reforms that may enhance occupational and sectoral mobility (Vermeulen et al., 2020). For one, there is the Danish policy of *flexicurity*, which seeks to (a) provide social security in combination with (b) training towards sustainable, upward mobility and (c) incentives for employers to hire employees lacking experience or skills (see Commission of the European Communities, 2007; Forge et al., 2010; Vermeulen et al., 2018; Vermeulen et al., 2020). Computer simulations show that whenever 'leaps' of the low-skilled newly unemployed to the high-skilled occupations in newly emerging sectors are unlikely, *vacancy chains* may emerge endogenously in which vacancies for higher skilled work are filled up by workers who thus leave vacancies that can be fulfilled by lower skilled workers (Vermeulen et al., 2020).

The second key mediating factor is that technological change affects countries' global competitiveness and employment in various ways. The productivity growth realised by adopting R&AI products in the applying sectors may lower unit costs and/or upgrade product propositions and thus effectively increase global demand. Moreover, innovative local R&AI making sectors (of which innovativeness may well be boosted by the local applying sectors) may also increase both local and global demand. This prospect warrants national innovation efforts and other ways to enhance absorptive capacity and stimulate adoption of R&AI. Moreover, technological change affects the distribution of production and service tasks in the applying sectors across countries. It may well be that R&AI substitutes workers in low-wage countries, accompanied by 'reshoring' work to other countries further down the production chain; however, given the many factors involved, such an outcome is by no means certain. Moreover, this substitution and resulting increase in unemployment may occur in low-wage countries, while the 'new' jobs in complementary tasks and the making sectors may emerge in high-wage countries. So, from the world-system perspective, and especially the current distribution of manufacturing work throughout global production chains, new technology may consolidate or even exacerbate international inequalities. This is to be considered a geographical realisation of Goos' polarisation. In this perspective, efforts of, for example, labour unions, to moderate the introduction of R&AI may (inadvertently) bring about a shift in the global competitive position.

Limits to Change?

Three primary variables drive employment and structural change in the direction of either one of the future scenarios.

The first is the rate at which the scope of applicability and therefore potential scale of substitution by R&AI technology changes. This scope and scale of substitution is limited not only by technological feasibility, but also by the research and development costs, the production and application costs, and, as such, the economic viability given wages. For instance, while it may be possible to replace a human performing tasks requiring high dexterity, it may be excessively costly. Moreover, the cost-economic viability of developing R&AI technologies for certain tasks is affected by the availability of workers able to do the tasks.

The second is the rate at which humans can take up new tasks and skills in new occupations in (potentially) emerging sectors. This depends on both the availability of labour (and notably, whether humans are tied up in existing sectors) and whether these workers can be educated, trained and up-skilled as fast as or faster than needed in the skills required. Moreover, humans must be educated and trained in the skills required for the tasks remaining for humans. Arguably, the work that remains requires creativity, intellectual flexibility, physical dexterity and social skills, each of which can only be taught to a limited extent. Indeed, there are limits to the *transferability* of certain skills, to the *educationability* of humans and, as such, to the dynamic efficiency of structural change, and these limits may very well apply particularly to creative work. Therefore, creativity as the next pivotal skill (after physical dexterity for the transition to an industrial society and knowledge processing for the transition to a service dominated economy), and as one of the last bastions of employment, may actually invalidate flexicurity. After all, whenever workers cannot be trained for the skills required, new sectors hinging on these skills will be hampered in taking off, while the flexibility to fire workers introduces a sustained pressure on social security.

The third is the rate and scale at which new employment opportunities emerge to compensate for the substitution and the productivity. This depends on the innovativeness of an economy, which is enhanced – according to Schumpeterian economists – by stimulating entrepreneurship. So, for one, future employment depends on educating and stimulating skilled innovators and entrepreneurs. Moreover, the (potential) employment opportunities thus created also need to be fulfilled, not only to create work for the unemployed, but also because doing so may well lead to further cascades of product and process innovations.

Given the structuralist interactions of countries, a rigid labour market may keep workers tied up and hamper the adoption of R&AI, which, therefore, may hurt global competitiveness of the applying sectors, prevent reaping new technological opportunities, and ultimately – ironically – cause unemployment. That said, dynamic efficiency of structural change requires both innovativeness and educational institutions. So, if creative work indeed is one of the last bastions, but creativity has a limited educationability, the labour market may become polarised with extensive income inequality.

CONCLUSION

In this chapter we have taken a comprehensive taxonomic structural change perspective in which we discerned labour economic effects within and across sectors, differentiated by *type of sectors* based on the relationship with R&AI technologies. Our findings provide support for claims that the creative industries are pivotal for modern economies (cf. Abbasi et al., 2017). Admittedly, further research should reveal the role and prominence of other emerging complementary tasks remaining after the adoption of R&AI, such as social or super-dexterous work. However, we assert that creative work may indeed emerge as one of the last bastions of employment, particularly in the (possibly extending range of) applying sectors – for example, in the form of embedded creatives.

Moreover, in general, countries that are able to retain employment in the making and applying sectors and realise labour mobility to exploit opportunities in the emerging sectors may also get the growth bonus in the form of spillovers, particularly in creative sectors known for non-homothetic demand kicking in with high incomes. In contrast, countries in which the applying sectors do not adopt (presumably productivity-enhancing) R&AI, lose global competitiveness and thereby demand, and also miss out on growing a local making sector catering to the applying sectors' needs, further cascades of innovation and so on. However, this also uncovers the requirements that market economies need to be innovative and dynamically efficient: opportunities must be created, R&AI and other productivity-enhancing technology must be adopted, and labour must be mobile and freed up flexibly. As such, several institutions in market economies (e.g., labour unions, social security, public education, innovation collaboration platforms in coordinated markets or, rather, liberal counterparts), as well as the current labour-market composition, moderate how the adoption of R&AI affects (the dynamics of) structural change.

It is stressed here that to understand the impact of technological change on structural change, three institutional drivers need to be regarded in conjunction: innovation to create technological opportunities, labour-market flexibility to enable freeing up labour by adoption of productivity-enhancing technologies, and education to enhance labour mobility to reap opportunities. Indeed, the institutional framework should be tuned to ensure the creation and retaining of a global competitive position in high-income, high-tech jobs in the making sectors and hard-to-automate complementary tasks in the applying sectors. Moreover, the institutional framework should accommodate the creation of innovative new sectors to employ the newly unemployed and increase labour mobility. With regard to sustainability of growth paths, of particular concern are low-wage countries doing low- to middle-skilled manufacturing work in the global production chain and developing countries with even fewer industrial capabilities. R&AI may well polarise labour markets and the current distribution of manufacturing work throughout global production chains, thus exacerbating inequality between countries. After all, these countries may fall behind because: (a) education cannot provide skilled labour for complementary tasks; (b) innovative capabilities are insufficient to reap opportunities in the making or emerging sectors; and (c) most spillover income consists of *local* demand and this is actually negatively affected.

An avenue for further research is investigating the impact of limited educationability for the emerging and supposedly increasingly prominent complementary tasks such as creativity, super-dexterity and social skills. Moreover, studies could examine which institutional framework is most conducive to relatively high employment and economic growth under limited educationability of these increasingly prominent tasks.

NOTES

1.   Ben Vermeulen and Andreas Pyka gratefully acknowledge EU H2020 funding, Grant 731726.
2.   Innovation Economics, Hohenheim University, b.vermeulen@uni-hohenheim.de.
3.   ORCID iD: 0000-0001-6207-6690.
4.   ORCID iD: 0000-0001-8102-5397.

REFERENCES

Abbasi, M., Vassilopoulou, P. and Stergioulas, L. (2017). Technology roadmap for the creative industries. *Creative Industries Journal*, **10**(1), 40–58.
Acemoglu, D. and Restrepo, P. (2016). The race between machine and man: Implications of technology for growth, factor shares and employment (NBER Working Paper). Rochester, NY: Social Science Electronic Publishing, Inc.

Acemoglu, D. and Restrepo, P. (2017). Robots and jobs: Evidence from US labor markets (NBER Working Paper). Rochester, NY: Social Science Electronic Publishing, Inc.

Acemoglu, D. and Restrepo, P. (2018). Artificial intelligence, automation and work (NBER Working Paper). Rochester, NY: Social Science Electronic Publishing, Inc.

Autor, D.H., Katz, L.F. and Kearney, M.S. (2006). The polarization of the US labor market (NBER Working Paper). Rochester, NY: Social Science Electronic Publishing, Inc.

Autor, D.H., Levy, F. and Murnane, R.J. (2003). The skill content of recent technological change: An empirical exploration. *Quarterly Journal of Economics*, **118**(4), 1279–333.

Baumol, W.J. (1967). Macroeconomics of unbalanced growth: The anatomy of urban crisis. *American Economic Review*, **57**(3), 415–26.

Brynjolfsson, E. and McAfee, A. (2011). *Race against the machine: How the digital revolution is accelerating innovation, driving productivity, and irreversibly transforming employment and the economy*. Lexington, MA: Digital Frontier Press.

Commission of the European Communities (2007). Towards common principles of flexicurity: More and better jobs through flexibility and security. 27 June. Retrieved from https://eur-lex.europa.eu/LexUriServ/LexUriServ.do?uri=COM:2007:0359:FIN:EN:PDF.

Cunningham, S. and Higgs, P. (2009). Measuring creative employment: Implications for innovation policy. *Innovation*, **11**(2), 190–200.

Dosi, G., Pereira, M.C., Roventini, A. and Virgillito, M.E. (2018). The effects of labour market reforms upon unemployment and income inequalities: An agent-based model. *Socio-Economic Review*, **16**(4), 687–720.

Fagerberg, J. (2000). Technological progress, structural change and productivity growth: A comparative study. *Structural Change and Economic Dynamics*, **11**(4), 393–411.

Fisher, A.G.B. (1939). Production, primary, secondary and tertiary. *Economic Record*, **15**(1), 24–38.

Ford, M. (2015). *The rise of the robots: Technology and the threat of mass unemployment*. Oxford: Oneworld Publications.

Forge, S., Blackman, C., Bogdanowicz, M. and Desruelle, P. (2010). *A helping hand for Europe: The competitive outlook for the EU robotics industry*. Luxembourg: European Union.

Frey, C.B. and Osborne, M. (2017). The future of employment: How susceptible are jobs to computerisation? *Technological Forecasting and Social Change*, **114**, 254–80.

Goldin, C. and Katz, L.F. (1998). The origins of technology: Skill complementarity. *Quarterly Journal of Economics*, *113*, 693–732.

Goldin, C. and Katz, L.F. (2007). The race between education and technology: The evolution of U.S. educational wage differentials, 1890 to 2005 (NBER Working Paper). Rochester, NY: Social Science Electronic Publishing, Inc.

Goos, M., Konings, J. and Vandeweyer, M. (2018). Local high-tech job multipliers in Europe. *Industrial and Corporate Change*, **27**(4), 639–55.

Goos, M., Manning, A. and Salomons, A. (2009). Job polarization in Europe. *American Economic Review*, **99**, 58–63.

Goos, M., Manning, A. and Salomons, A. (2014). Explaining job polarization: Routine-biased technological change and offshoring. *American Economic Review*, **104**(8), 2509–526.

Griliches, Z. (1969). Capital: Skill complementarity. *Review of Economics and Statistics*, **51**, 465–68.

Hall, P.A. and Soskice, D.W. (2001). *Varieties of capitalism: The institutional foundations of comparative advantage*. New York, NY: Oxford University Press.

Hearn, G.N. and Bridgstock, R.S. (2014). The curious case of the embedded creative: Creative cultural occupations outside the creative industries. In C. Bilton and S. Cummings (eds), *Handbook of management and creativity* (pp. 39–56). Cheltenham, UK and Northampton, MA, USA: Edward Elgar Publishing.

Keynes, J.M. (1930). *A treatise on money: The applied theory of money*. AMS Press.

Kuznets, S. (1966). *Modern economic growth*. New Haven, CT: Yale University Press.

Leontief, W. (1983). Technological advance, economic growth, and the distribution of income. *Population and Development Review*, **9**(3), 403–10.

MacCrory, F., Westerman, G., Alhammadi, Y. and Brynjolfsson, E. (2014). Racing with and against the machine: Changes in occupational skill composition in an era of rapid technological advance. *Proceedings of the Thirty Fifth International Conference on Information Systems*. Auckland, New Zealand.

Mokyr, J., Vickers, C. and Ziebarth, N.L. (2015). The history of technological anxiety and the future of economic growth: Is this time different? *The Journal of Economic Perspectives*, **29**(3), 31–50.

Moretti, E. (2010). Local multipliers. *American Economic Review*, **100**(2), 373–7.

Napoli, P.M. (2014). On automation in media industries: Integrating algorithmic media production into media industries scholarship. *Media Industries Journal*, **1**(1).

Pasinetti, L. (1981). *Structural change and economic growth: A theoretical essay on the dynamics of the wealth of nations*. Cambridge, UK: Cambridge University Press.

Saviotti, P.P. and Pyka, A. (2004). Economic development, qualitative change and employment creation. *Structural Change and Economic Dynamics*, **15**, 265–87.

Saviotti, P.P. and Pyka, A. (2008). Product variety, competition and economic growth. *Journal of Evolutionary Economics*, **18**, 323–47.

Schumpeter, J.A. (1942). *Capitalism, socialism and democracy*. Harper & Row.

Silva, E.G. and Teixeira, A.A. (2008). Surveying structural change: Seminal contributions and a bibliometric account. *Structural Change and Economic Dynamics*, **19**(4), 273–300.

Vermeulen, B., Kesselhut, J., Pyka, A., Saviotti, P.-P. (2018). The impact of automation on employment: Just the usual structural change? *Sustainability*, **10**(5), 1661.

Vermeulen, B., Pyka, A. and Saviotti, P.-P. (2020). Robots, structural change and employment: Future scenarios. In K.F. Zimmermann (ed.), *Handbook of labor, human resources and population economics*. Cham, Switzerland: Springer Nature.

Vivarelli, M. (2007). Innovation and employment: Technological unemployment is not inevitable – some innovation creates jobs, and some job destruction can be avoided (IZA Technical Report). Bonn: IZA.

Vivarelli, M. (2013). Innovation, employment, and skills in advanced and developing countries: A survey of the economic literature. *Journal of Economic Issues*, **48**, 123–54.

Ward, W.H. (1967). The sailing ship effect. *Physics Bulletin*, **18**(6), 169.

PART II

Digital disruption and creative work

5. New economic infrastructures for creative work[1]

Ellie Rennie[2] and Jason Potts[3]

INTRODUCTION

The number of people working in the creative services sector is growing faster than the overall Australian workforce, yet the income of cultural producers – artists, musicians, performers and screen producers – is in decline (Digital Media Research Centre (DMRC), 2018; Throsby and Petetskaya, 2017). Lawful consumption of digital content is growing (Department of Communication and Arts (DOCA), 2018), yet many creative practitioners are not seeing appropriate returns from these sales.

The reasons that creative practitioners are facing these challenges are complex. The economic dominance of online platforms – in particular, streaming services and social media – is increasingly impacting how some creative practitioners reach audiences, requiring them to adapt what they do and how they allocate their time. Those whose practice involves fewer touchpoints with digital platforms can nonetheless find themselves in difficult circumstances when it comes to intellectual property and contracts (such as a musician not understanding the copyright implications of sampling, or a screen producer managing tax offsets).

Over the past decade, technology companies have become powerful players in the creative economy. Streaming services, online stores and social media platforms make it easy for audiences to receive and pay for content, to follow a creative practitioner's work and to express their preferences. In theory, these same attributes should also benefit creative practitioners. However, Australia's creative and cultural producers are generally not seeing greater financial returns for their work, or finding the business aspects of their work any easier.

Distributed ledger technology, also known as *blockchain*, is emerging as one way to rebalance the cultural economy in favour of creative practitioners. In theory, industries that are reliant on digital payments (especially micro-transactions and when contracting between parties can be complex) will stand the most to gain from the arrival of blockchain technology. In addition, the ability to authenticate a work as it passes from one buyer to the next, and to generate unique digital works, should benefit industries in which scarcity is valued. Creative-industries blockchain platforms are experimenting with royalty payments for music and screen works, proving the authenticity of visual artworks and fashion, avoiding ticket scalping and more. For creative practitioners, this should mean simpler and more transparent transactions, easier contracting, fewer overheads and less reliance on intermediaries. Streamlined processes for collaboration between creative practitioners might also emerge.

However, the embryonic blockchain-enabled creative economy has a difficult road ahead. Neither old industry incumbents nor new technology platforms have demonstrated a willingness to embrace an open and accessible *internet of value* (as blockchain is known). Without concerted efforts to coordinate practitioners and stakeholders (arts organisations, creative firms, funding bodies, collecting societies and others), including shared digital infrastructures and open standards, these benefits may never be realised. One approach to overcoming these barriers is public provision of digital-platform infrastructure, or what former CSIRO CEO Adrian Turner (2018) calls *industry utilities*. In this chapter, we consider what such an approach would mean for creative work.

CREATIVE INDUSTRIES TODAY

The creative industries can be understood in terms of their contribution to the economy, and for their non-economic value, such as the role the cultural sector plays in nationhood, wellbeing and informing our understanding of the world (Bakhshi and Cunningham, 2016; Keat, 2000). Fair and efficient methods for creative practitioners to produce and share their work can result in multiple societal benefits.

In 2016, there were 347 190 people working in creative occupations in Australia (using the definition of "creatives" from the United Kingdom (UK) foundation Nesta, as outlined in Higgs and Lennon, 2014). Of these, 162 170 were employed in the creative industries (made up of music, visual and performing arts; film, television and radio; advertising and marketing; architecture and design; software and digital content; and publishing), with the remainder doing creative work in other industries (DMRC, 2018). When creative occupations and non-creative support roles in the creative industries are combined, the size of Australia's creative workforce was 593 840 in 2016 (DMRC, 2018).

Those who work in creative occupations in the creative industries are likely to be highly networked sole-traders or running micro-businesses. As they are often self-employed or employed on short-term contracts, they must either learn to manage their career and business, or partner with others who are able to manage business tasks for them (Bridgstock, 2013).

POLICY RESPONSES TO DATE

Creative-industries policies have so far focused on digital rights reforms, skills and the transformation of existing content to new means of discovery and consumption of creative works. Researchers have also identified a need to educate creative workers to be more entrepreneurial and to deal with the business aspects of their work (Australian Film Television and Radio School, 2019; Bridgstock, 2013; Flew, 2012). Nesta has concluded that nations need to invest in more than broadband pipes in order to compete in the creative economy; "it also calls for investment in new skills, the re-engineering of production processes and new business models to create and capture value" (Bakhshi et al., 2013, p. 12).

Creative-industries policy is now beginning to consider the full impacts of platforms on culture. For instance, a UNESCO (2018) report asserts that governments need to develop policies and infrastructures to deal with the cultural outcomes of the platform economy:

> Few countries have been able to design comprehensive agendas or provide adequate infrastructure to deal with the transformation in the digital domain from a linear or pipeline model to a network configuration. There is a danger that the public sector will lose its agency on the creative scene if it remains unable to address challenges such as the rise and market concentration of large platforms, the unfair remuneration of artists or the monopoly on artificial intelligence. (p. 21)

Broadcasters in the UK have written an open letter calling for laws to be enacted that ensure public service broadcasting content remains prominent on screens out of concern that "[g]lobal technology players have growing influence on what UK audiences discover when they turn on their screens" (McCall et al., 2018, para. 5). In particular, these broadcasters are concerned that video on demand (VOD) services, including subscription video on demand (SVOD) platforms and social media, recommend programmes based on either search algorithms or paid promotion of channels and content.

CHALLENGES FACING CREATIVE PRACTITIONERS

Employment in the creative services grew at a rate of three times that of the Australian workforce in 2011–15 (DMRC, 2018). Advertising and marketing, architecture and design, and software and digital content experienced the highest rates of growth. During the same period, the mean income of creative practitioners working in screen, music, performance, publishing and visual art fell (DMRC, 2018).

Research shows that many creative workers believe that spending more time on creative work and improving skills will lead to greater success, but finding the time is difficult (Throsby and Petetskaya, 2017). Creative practitioners also accrue costs in relation to access to specialised knowledge, up-skilling, administrative processes and enforcement. For instance, research from Screen Australia found that while the Producers Offset is beneficial for the screen industries, producers have incurred higher legal and administrative costs as a result, including taking on new staff or contracting specialists (Screen Australia, 2012; 2017).[4] The high costs of formal contracting can result in producers accepting informal arrangements, resulting in weak organisational infrastructure and support, misallocation of resources and difficulty in proving viability for finance. The result is an increase in risk and uncertainty for some producers.

The Screen Industry

Technological advances in home entertainment and online distribution have profoundly disrupted the screen industry. VOD platforms enable audiences to watch almost any content, at any time, on a variety of screens. Digitally empowered audiences are splitting their viewing across multiple platforms, including VOD, cinema, broadcast and subscription television, and home entertainment. Audience choices are a boon for consumers, who now expect easy (if not free) access to high-quality content, and digital technology provides new tools for screen producers to reach audiences; however, digital disruption and audience fragmentation have challenged all aspects of the screen industry. Audience expectations are high and competition for viewers is rising, at the same time as the business structures that have traditionally financed Australian content are profoundly challenged.

Independent film
Digital disruption is acutely affecting the business model that historically underpinned independent film, where the initial experience has often been that "analogue dollars have been traded for digital cents" (Screen Australia, 2015). Australian cinemas are widely attended, but many independent films, including many Australian films, are struggling to reach cinematic audiences. The number of films released in Australia has more than doubled in the last 10 years (Screen Australia, 2018). Increasing numbers of independent films are competing for a fairly static share of the box office that is earned by films with relatively small releases, while big-budget "blockbusters", which open on a larger number of screens and have high global awareness, have doubled their

share of the Australian yearly box office over the last decade (Screen Australia, 2018). As cinema audiences increasingly prefer bigger-budget films, and as cinemas offer more independent films from foreign markets such as Asia, Australian films can struggle to cut through.

Television
While Australians are consuming more video content than ever before and the television is still the preferred device to watch, this device is increasingly being used for content other than broadcast content, particularly by younger Australians (Australian Communications and Media Authority (ACMA), 2019). At the end of June 2019, 55 per cent of Australian households subscribed to an SVOD service, and in May 2019, 11 million Australians had access to Netflix in their home, while games and social media platforms provide further options that are particularly popular with young audiences (ACMA, 2019; Roy Morgan, 2019; Telsyte, 2019). This increased competition is changing broad-caster commissioning and providing new platforms for producers.

When streaming services such as Netflix purchase content from producers, they sometimes buy exclusive rights, for all territories, in perpetuity (McCabe, 2019). While this model is not common across the industry, it provides two challenges. First, it cuts off the long tail of sales that have provided production companies with the means to develop new works. Australian films overwhelm-ingly do not recuperate costs from their initial release but earn income over time (George and Rheinberger, 2017). Moreover, only streaming services have access to the fine-grained information about consumer behaviour generated through their platforms and social media, which informs these services' com-missioning processes.

Social media
These platforms are the main distribution outlet for some emerging producers and those catering to new, mostly young, audiences (Cunningham and Craig, 2019). While platforms can provide powerful audience access and insights to creative producers, producers have little power to alter their terms of engage-ment with platforms (although there is some choice between platforms). Social platforms also may not provide suitable financing and distribution opportuni-ties for content such as higher-budget drama.

The Music Industry

The music industry is complex, both in terms of the flow of money and its interaction with digital platforms. Innovations in reproduction, recording and printing have restructured the music market throughout its history (Dommann, 2019). When the internet arrived, the music industry was the first to experience major disruption (via file-sharing platforms such as Napster), making it a key battleground for digital rights management. The launch of subscription music platforms has altered dynamics once again.

In Australia, the arrival of music-streaming platforms has aligned with an increase in lawful consumption of music. While illegal downloading of music still occurs, a DOCA (2018) consumer survey on online copyright infringement found that the percentage of Australians lawfully listening to music online has increased: 91 per cent of Australians now pay for some music online compared with 52 per cent in 2015. Consumer spending on music in the United States (US) is now back at an all-time high (CitiGPS, 2018) due to the accessibility of music via streaming services and smartphones. ARIA's 2018 music industry figures indicated a 12.6 per cent annual growth in music revenue, with streaming revenue accounting for 71.4 per cent of the overall market by value, a growth of 41.2 per cent over the year. Physical products accounted for 15 per cent of the total market, with sales from vinyl increasing for the eighth consecutive year (making up just under 28 per cent of revenues from physical formats) (ARIA, 2019).

The rise of streaming platforms has also brought with it new intermediaries and challenges. Spotify curators (in-house, independent or algorithmic) select songs that make it onto playlists with sometimes millions of followers. Although Spotify does not allow pay-per-play services, some third parties have been said to accept "playlist payola" for independent playlists (Resnikoff, 2016). Spotify abandoned an experiment in 2019 in which it allowed independent artists to upload their own music to the platform, thereby requiring that artists go through an approved distributor if they wished to be on the platform (Spangler, 2019). And while money is being made on Spotify, it is funnelled through record labels, publishers and collecting societies, leaving artists with small returns. A 2015 report from the Berklee Institute for Creative Entrepreneurship (Boston, MA) found that musicians needed to be "knowledgeable and vigilant" when it came to being rewarded for their work because "[f]aster release cycles, proliferating online services, and creative licensing structures make finances and revenue even more complex to understand and manage" (Rethink Music, 2015, p. 3).

In a manner similar to radio, streaming services allow audiences to listen to music without having to purchase it. Unlike in the radio era, however, streaming services do not entice consumers to buy music (to listen to at any time) because songs remain accessible on demand. Rights models and contracts do not reflect these affordances, so many artists make less from streaming than they did when audiences purchased physical copies (CitiGPS, 2018; Donoughue, 2017). In addition, while digital platforms should be able to provide greater transparency when it comes to use and payments, information arrives in artists' hands "as a hefty stack of paper" rather than as real-time useful data (Rethink

Music, 2015). A number of international reports and inquiries have concluded that there are few incentives for those holding royalty money to pay it to artists (O'Dair, 2019). Problems in data management have resulted in what is known as the *black box*: a reservoir of royalties that are not paid to artists because they cannot be traced. In response, businesses have emerged that trace royalties for artists by processing large datasets (such as Sound Exchange). In at least one instance, a music label has negotiated an advance deal with streaming services, yet has kept payments that cannot be traced due to data management problems (Singleton, 2015).

Changes in technology and audience behaviour are also influencing the business management model for music. Music companies are diversifying, with multiple divisions under one roof, including managing production, publishing, marketing, live performances and online distribution. Platforms such as Band Camp are assisting smaller record labels to achieve the full scope of services that were once only available through larger labels. Beyond this, CitiGPS (2018) predicts "organic forms of vertical integration" (p. 3), whereby companies that own online distribution sites "morph into music labels" (p. 3), providing business support for artists while capturing profits held by current intermediaries such as record labels and concert promoters.

Investments and Donations

In the late 2000s, crowdfunding platforms were heralded as a way to democratise arts funding (Boeuf et al., 2014; Brabham, 2017). For some artists, they have been a useful vehicle to market works and receive donations, particularly from their immediate networks. For others, the work required to launch and manage a successful crowdfunding campaign, including rewards for donations, can outweigh the benefits.

Newer platforms such as GigRev, Music Glue and Patreon overcome this somewhat by enabling artists to create an ongoing connection to fans and build a dedicated support stream. For instance, on Patreon, a podcaster or YouTube producer can provide additional episodes and merchandise in return for a monthly subscription. These platforms work on the basis that "Consumers don't want to support Spotify or Netflix, they want to support the artists" (Kevin Brown of GigRev, as cited in CitiGPS, 2018, p. 77). Crowdfunding platforms have shown that it may be possible for artists to seek investment from a larger pool of smaller funders (who receive an equity stake in a work).

THE OPPORTUNITY OF BLOCKCHAIN

Blockchains are a means for people or machines to move value across the internet (Rauchs et al., 2018; Werbach, 2018). The "value" that people or machines are moving when they use a blockchain might be money (in the form of a digital currency) or it could be a token representing an equity stake in an asset, or a unique digital object (such as a digital artwork). Importantly, blockchains ensure that the same value cannot be copied or spent twice. Blockchains do this by creating a ledger of transactions, which shows when exchange has occurred, updating the shared record to reflect who owns what. The technology provides a trustworthy means of exchange, so it is said to reduce the need for intermediaries. For example, property sales that use blockchain overcome the need for conveyancers (such as the US-based platform Meridio).

Smart contracts are a means of automating agreements between parties using blockchains. A common form of smart contract in use today is sending cryptocurrency to an escrow account and instructing that account (using software) to automatically pay the funds to a purchaser when an item has been received. The significance of smart contracts is that they can make it cheap, easy and safe to send and receive payments. In addition, they can contain specific instructions (such as paying all rights holders their agreed portion of payment every time a work is purchased) and can be coded to comply with laws and regulation. For instance, in the visual arts, smart contracts can be used to ensure that artists automatically receive re-sale royalties each time an artwork is sold (in Australia this currently occurs through manual processes and only applies to works over AU$1000).

In this section, we provide an overview of the many different ways that blockchain can potentially be applied to the creative industries, using examples of platforms that have been proposed or trialled. Some of the examples provided might have already failed or disappeared, but nonetheless provide insight into the aspirations of developers and potential uses of the technology.

What are Blockchains?

Fundamentally, blockchains provide an economic infrastructure that enables parties to coordinate among themselves. While there are other systems for payments and exchange (e.g., PayPal, eBay), blockchains do not require a central authority to store and maintain records, which can cause inefficiencies, concentrate power and provide opportunities for manipulation. Blockchains are secure ledgers within which entries are entered and validated, but the contents of the ledger are agreed upon by a distributed system of computers, rather than a firm or government. In the case of Bitcoin, for instance, these entries have the qualities of money (scarce, fungible, unit of account) and can be used as money without the need for banks. The same processes can be used to record other forms of property, including intellectual property.

Different blockchains have different properties. Public blockchains (including Bitcoin and Ethereum) allow anyone who has the required hardware, software and electricity to participate in keeping the network decentralised and secure (known as "mining"). Private blockchains (such as Hyperledger and Quorum) are used in areas that require fast transactions, when fees for transactions are not desirable, or when a particular group may wish to retain control over aspects of the data. It is possible that the blockchain economy will evolve into linked, interoperable blockchains, each with its own features.

Blockchains are already being used to coordinate supply chains that operate across multiple phases of trade and contracting. For the creative industries, the ability to trace a work across a supply chain means it can be used to establish genuine articles from copies or fakes. For instance, in March 2019, it was reported that luxury brand owner LVMH, owner of Louis Vuitton, would launch a platform for verifying provenance (Allison, 2019). Platforms such as this will reduce the need for verifying and auditing information about the quality and characteristics of a product as it moves along a supply chain or in any context of joint production. This may include not only provenance, but also contracting and subcontracting, licensing, certification, credentialing, royalties or broadly any context that requires a trusted third party to verify or authenticate information that is an input into production or consumption.

Smart contracts and payments

A "smart contract" is a piece of code that automates the transfer of digital assets between parties when predetermined conditions have been met. Ethereum, one of the largest blockchains in terms of number of transactions, is sometimes described *as programmable money* in that it can execute a transaction under specified events, akin to a vending machine that only dispenses an order once the money has been received (Szabo, 1997). Smart contracts can also be used to automate rights payments and terms.

For the creative industries, there are clear benefits to making contracts and payments more efficient. Smart contracts can be used to facilitate digital collaboration by clearly ascribing, attributing and remunerating work. Smart contracts can also be used to automatically remunerate artists a predefined percentage or amount on secondary sales. Artists can create agreements with individual contributors to a song (e.g., songwriter, band members). An agreed percentage of the total revenue is paid to contributors based on how many streams were recorded on the blockchain. Translating these and other processes and law into code is a difficult task, and is unlikely to be done retrospectively for existing works. However, for creative practitioners who are in control of their works and distribution (such as an unsigned musician), this may be an attractive and viable option.

For instance, a filmmaker typically negotiates and manages numerous agreements in the process of getting a film to market. Even after the production has been released in cinemas, there will be ongoing processes of payments, and screenings on other platforms and overseas. Other industries with similarly complex legal processes, such as real estate, are already using smart contracts (Pettit et al., 2018). These contracts are developed by legal experts and mirror non-digital contracts. The difference is that they can be programmed to carry out processes under certain conditions. A smart contract in the film industry might update when there are changes to rights-holder agreements, when a licensee fee is paid and when a sale has been made into a territory, and might automate payments when gross receipts have been achieved. This allows payments down the "revenue waterfall" to be approved and processed automatically by the rights-holder in real time under agreed terms.

Royalty payments

Music generates millions of micro-transactions a day, which need to be distributed to multiple stakeholders. Artists typically use collecting societies to receive royalties from Australian use of works, and rely on partnerships between Australian and international collecting societies to receive revenue from overseas use. Worldwide, collecting societies are struggling to manage performance rights data due to the volume of data and inconsistencies in how data is reported and handled (Rethink Music, 2015). In 2017, APRA-AMCOS stated that it had seen a ten-fold increase in the amount of data received and needing to be processed over a seven-year timeframe (APRA-AMCOS, 2017). Some predict that collecting societies will soon outsource the management of performance rights to tech companies (van Rijn, 2018).

Collecting societies have begun investigating the use of blockchain technology for royalty tracking and timely payments. Australia's Copyright Agency Limited, for instance, instigated a pilot to test how blockchain could be used to administer re-sale royalties of visual artworks, commencing with Indigenous artworks (Copyright Agency, 2018). Another group working together to this end comprises IBM and blockchain platform Hyperledger Fabric, which have partnered with US collecting society the American Society for Composers, Authors and Publishers (ASCAP), France's Society of Authors, Composers and Publishers of Music (ACEM), and the UK's PRS for Music (Allison, 2017). This large group is reported to have come together "to model a new system for managing the links between music recordings International Standard Recording Codes (ISRCs) and music work International Standard Work Codes (ISWCs)" (para. 3) and tackle a "long-standing issue" with metadata and authoritative copyright data (Allison, 2017).

The Open Music Initiative (Berklee School of Music and MIT) has been developing standards for technology platforms to use via application programming interfaces (APIs) and has attempted to bring industry together to adopt shared systems. Currently, songs can be assigned an ISRC, which is an identification system for sound recordings. The ISWC is similar except that it applies to compositions. These codes are not linked, which means that composers do not always see returns from recordings of their music. In addition, there is no common method of working with these codes, meaning that they end up needing to be re-entered into different databases, resulting in lost payments.

O'Dair (2019) argues that blockchain may succeed where previous attempts at a single database of rights information have failed (such as the Global Repertoire Database; see Cooke, 2014) because blockchains align incentives, meaning that different entities can act according to their own self-interest, with outcomes that benefit all (O'Dair, 2019). However, some pioneers in the field are less positive. George Howard from the Berklee College of Music argues that entrenched interests and problems in metadata may stifle industry coordination. Instead, the Open Music Initiative is creating a platform for new "controlled compositions", which allow the same individual to own the composition and performance rights (Howard, 2018; 2019).

Rare digital objects

Digital art has mostly failed to generate value for creators because it is not seen as rare; it can be copied or manipulated in ways that undermine the principles of scarcity and originality that drive art markets (O'Dwyer, 2018). Blockchain addresses this through what are commonly referred to as "rare digital objects" or "digital collectibles", made possible through a software innovation called *non-fungible tokens* (NFTs).

In economic terms, fungibility is a characteristic of goods or commodities whose individual units are equivalent, interchangeable and indistinguishable from other units of equal value. Fungible assets include money, bonds and precious metals. When an asset is non-fungible, it is unique and non-interchangeable. NFTs represent unique physical or digital assets on a blockchain. NFTs secure and prove asset scarcity and individuality. As such, blockchain-enabled NFTs facilitate asset provenance or tracking, and verify asset ownership or authenticity.

FUTURE SCENARIOS FOR A BLOCKCHAIN-ENABLED CREATIVE ECONOMY

So far, we have considered a range of creative-industries blockchain experiments, as well as the industry context they seek to disrupt. In this section, we define the blockchain creative economy and consider different scenarios for how it could evolve.

Blockchains coordinate economic activity by providing a mechanism to reach agreement about economic facts (Novak et al., 2018). Aspects of market capitalism can be done on blockchains: money and value; identity and asset registries; property rights; exchange mechanisms; law; finance; and govern-ance through voting. Blockchains can therefore be described as an economic infrastructure that enables parties to coordinate, performing a similar role to institutions (e.g., the banking system, laws).

One significant benefit of blockchain technology is that it can make it easier, faster and cheaper to perform certain tasks (lowering what economists call *transaction costs*). For instance, when verification occurs through software, the need for costly intermediation and auditing is reduced or removed. Catalini and Gans (2017) argue that blockchain adoption lowers the costs of verifica-tion and helps prevent opportunism, and that this will increase the efficiency and scope of markets. Blockchain is therefore an important tool for the creation of direct, decentralised peer-to-peer systems.

A Decentralised Economy Versus a Private Blockchain Economy

There are two different models for the design and use of blockchain tech-nology: public blockchains and private blockchains. The difference between a public blockchain and a private blockchain model relates to how governance occurs – whether it is open to anyone to participate, or whether it is restricted to a selected group.

One example of private blockchain development is occurring under the banner of Enterprise Ethereum. Based on the public Ethereum blockchain code base, Enterprise Ethereum enables companies to create private block-chain environments where only "permissioned" entities participate (such as banks participating in JP Morgan's Quorum, which is currently being used for interbank information). This scenario is attractive to those who wish to control privacy features, transaction speed or which parties have access to the ledger. The Enterprise Ethereum Alliance is developing standards and specifications that will allow private blockchains that use Enterprise Ethereum to talk to and work with each other. Importantly, private blockchains are not *trustless*: for example, it is easier for a group of defined actors to conspire to manipulate the ledger, making these blockchains reliant on non-technological checks and balances (laws, regulations and codes of practice).

In contrast, anyone can participate in running a public blockchain such as Bitcoin or Ethereum and earn rewards for participation. If changes to the code are proposed by developers, those running the software must make a choice whether to accept or reject the change. In a public blockchain, the applications that are built on the blockchain perform certain tasks as determined by the developers, such as a marketplace for music or art. Those who endorse public blockchains over private blockchains argue that the tools and applications in a public blockchain environment are more fluid than the current operations of firms and hyper-responsive free market dynamics, and may involve decentralised decision-making (see ConsenSys, 2018). Public blockchain applications are often peer-to-peer environments by design and do not need intermediaries, whereas private blockchains are constructed to complement or perform work for intermediaries and firms.

In 2017, Spotify acquired MediaChain Labs, a company that was building a distributed blockchain for royalties settlements that embedded timestamps and metadata into music files (McIntyre, 2017). Spotify has not revealed how or whether it is using the technology. However, the company announced in June 2019 that it is joining the Libra blockchain (initiated by Facebook) for consumer purchasing of music, with Alex Norström, Spotify's Chief Premium Business Officer, claiming it is "an opportunity to better reach Spotify's total addressable market, eliminate friction and enable payments in mass scale" (Spotify, 2019, para. 5).

Competitive private delivery of administrative infrastructure – whether built on a public or private blockchain – is highly desirable with respect to speed of development, but is not necessarily the solution from the perspective of the creative industries. The main problem, as already indicated above, is that, if successful, these businesses, as platforms, have strong natural monopoly characteristics that will be able to be competitively exploited unless subsequent mechanisms, such as regulation, are enacted.

A key problem, which can readily be seen in the current competitive positions of recent generations of digital-platform technologies – such as search (e.g., Google), social media (e.g., Facebook, Twitter) and digital marketplaces (e.g., Amazon, Alibaba) – is algorithmic inscrutability. In essence, it can be very difficult to provide public accountability and audit control under such conditions.

An Industry Utility Approach with Public Support

A *public utility* is an organisation that provides economic infrastructure that has natural monopoly characteristics owing to large upfront or fixed costs and for which economic efficiency considerations require either public ownership or public regulation (if privately owned). An industry utility is a public utility that provides targeted infrastructure for a particular sector.

In this section, we consider what an industry utility for the creative industries might look like. We call this hypothetical utility the *Australian Creative Blockchain*. Further research is needed to determine the feasibility of an Australian Creative Blockchain, how it might be piloted and staged, and its potential benefit to other cultural organisations.

An Australian Creative Blockchain could include a registry of works and contracts, and provide a basis for automated payments. Important elements are:

- smart contracts for intellectual property that reduce the cost of existing contractual processes, maintain definitive records of rights and distribution arrangements, and enable new business models;
- a means to administer grants and attract investment in the arts and screen industries;
- automated payments from various points in the system (theatres, platforms, audiences) to rights holders, assisting artists and producers to recover royalties more efficiently, especially across international borders.

How would this unfold? The Open Music Initiative from Berklee School of Music and MIT is commencing with student-produced music (where the individual owns the composition and recording rights) and providing a platform that would make it easy for filmmakers to license these works for use in productions using smart contracts. Such a model could be explored in Australia, consisting of a platform for intellectual property registries and smart contracts that are tailored to Australian regulation and law. Some of Australia's cultural agencies support musicians through grants for creating artistic works, mentoring and touring. These works might be administered through the Australian Creative Blockchain, thereby producing metadata that would assist cultural institutions to better assist artists. Further down the track, this infrastructure could provide an opportunity to connect into industry-led blockchain innovations in content distribution, ticketing, investment and engagement. In addition, cultural institutions could play a part by understanding and working with alternative business models (such as cooperative models) that are made possible by the technology.

Ultimately, an Australian Creative Blockchain could be joined to creative agencies and arts councils in other countries (e.g., Creative New Zealand, National Arts Council Singapore) or to creative or cultural digital-platform utilities as they emerge in other countries or regional jurisdictions.

The case for public provision or commissioning of an Australian Creative Blockchain as an industry utility is similar to the case for other business or trade infrastructure, such as electricity grids or communications networks, ports and roads, standards, or commercial law. An Australian Creative Blockchain has the potential to furnish significant net economic benefit, but has public good and natural monopoly characteristics.

First, such assets have *public good* characteristics due to market failure in consequence of fixed costs of development, as well as *natural monopoly* characteristics, owing to the interaction of increasing returns to scale and competitive pricing. Economic efficiency considerations imply collective provision, whether as provided by government or through an industry consortium (in economics, this is called a *club good*). The argument for government rather than private consortia provision of an Australian Creative Blockchain rests on the consumer benefits of competitive open access to the infrastructure. A private consortium could potentially block access to competitors, and thus, attain monopoly rents and other forms of market control from doing so. However, those same rents are also the incentive for private research and development, financing, and construction costs to build an Australian Creative Blockchain by a private consortium.

Second, such an infrastructure may not make a profit, but will likely enhance the overall productivity of the creative industries. Automating many routine aspects of creative-industries business administration using blockchain-enabled technologies could potentially lower operational business costs and support the substitution of creative labour away from administrative tasks and toward creative content production. In both dimensions, this pushes creative businesses toward improved profitability and growth, and therefore, sustainability.

As a payments, contracting and administrative infrastructure utility, the automation brought by an Australian Creative Blockchain could increase business formality and compliance. This characteristic would improve the prospects of creative-industries businesses in attaining:

- cash-flow management through payments technologies;
- legal assurance through more effective contracting;
- finance to enable growth and investment;
- insurance for risk management;
- trade and export opportunities through property rights management and tax law;
- domestic and international tax, trade, and regulatory compliance.

Moreover, whichever country or region is able to develop a protocol and infrastructure for the business administration of creative industries may become the standard adopted by other countries. Such an outcome would afford its own home sector a strategic advantage in global trade in the export of creative content, and an effective tool of soft power and the export of administration expertise.

Nevertheless, these are complex issues that are fraught with substantial technological and market uncertainties, and are dependent upon the continuous progress of industrial-scale innovation in the development of blockchain technology and its supporting industrial ecosystem. We therefore recommend further study and economic analysis of the optimal approach to public finance, the optimal social investment and the scope of regulation of a creative-industries utility.

CONCLUSION

Business administration in the creative industries can be difficult and time-intensive, and this imposes significant costs on creative-industries business. Overall, these costs constrain the economic development of the sector, and may limit the creative works that are produced and made available to the public.

Creative practitioners' income and time are dependent, not just on their own business acumen, but also on factors beyond their control, including broader industry transformations that affect audience and buyer behaviour. While older distribution systems (including broadcast television and cinema) remain a source of income for some, digital platforms are changing how works are commissioned and how income is derived in parts of the creative industries. Some creative practitioners are able to navigate newer environments successfully, while others are left behind.

Blockchain is an economic infrastructure that may overcome some of these challenges. In essence, any content that can be digitally produced or represented in the initial instance can be written to a blockchain, establishing a secure trusted record of creation that can be subsequently managed, traded, licensed, tracked, verified, permissioned, and so on. This establishes authenticity and priority (as a rights register for intellectual property does). In addition, blockchain creates an integrated native trading and value-transfer platform (as a payment and banking system does), a legal contracting and enforcement mechanism (as a court system does) and a metadata-based search and matching platform (as a search engine and internet market does), and facilitates pooling and governance of resources (as a company structure does).

By integrating these distinct functions onto a single technological infrastructure, blockchain may significantly change how the creative industries operate on the business side, and may enable peer-to-peer markets. As a result of these features, creative practitioners may be empowered to find new means of connecting with audiences and fans, and may find that creative collaboration is easier to manage.

Blockchain technology may enable a more efficient and transparent creative economy. Peer-to-peer platforms hold significant promise by allowing creative practitioners to deal directly with fans. In addition, aspects of the creative economy that require coordination between practitioners and stakeholders (arts organisations, creative firms, funding bodies, collecting societies and others) may benefit from shared digital infrastructures and open standards.

One approach to such coordination is an industry utility for Australia's creative industries that would provide tools and partnerships that use shared, distributed technologies. The potential outcomes are efficiency, fairer systems and new models for grants and investment.

Creative practitioners who earn income from their work are likely to do better if administrative and operational costs are reduced. For "complex creative cultural goods" in particular, such as those involving teams and production schedules (Caves, 2002), the legal, resourcing and finance aspects of creative-industries business operations reduce time and effort devoted to actual creative production, and can limit the growth of creative businesses. This vulnerability is particularly acute in the domain of new and small businesses, contractors, part-timers, semi-professionals, and those seeking to expand and grow in the formal sector.

Existing cost burdens have not been alleviated with the availability of digital production and distribution (the forces of "disintermediation"). As the platform economy grows via streaming services – used in a combination of algorithm-driven "smart" receivers (televisions, speakers) – it is possible that creative practitioners will increasingly have to deal with opaque systems and platforms. Early stage distributed and automated systems (including blockchain applications) may resolve some of the current complexities of the creative economy. However, the same technologies encoded into private platforms can be used to the benefit of some over others.

NOTES

1. This chapter is an extract of a Provocation Paper titled "Blockchain and the creative industries", which was prepared at the request of the Australia Council for the Arts, Screen Australia and the Australian Film, Television and Radio School. It was published in December 2019 by the Blockchain Innovation Hub at RMIT University and co-authored with Ana Pochesneva (who worked extensively on the Appendix of the Provocation Paper, not included here). We would like to thank the Australia Council for the Arts, Screen Australia and the Australian Film Television and Radio School for their advice and feedback. In particular, Paul Mason, Rebecca Mostyn, Rachel Perry, Neil Peplow, Georgie McClean, Tessa Sloane and Patrick May provided key insights during a workshop, as well as thoughtful comments and advice throughout the drafting of the Provocation Paper. Additional thanks to Jay Mogis, Ben Morgan and Chris Berg.
2. ORCID iD: 0000-0001-6792-9744. Media and Communication, RMIT, ellie. rennie@rmit.edu.au.
3. ORCID iD: 0000-0003-1468-870X.
4. While the research found that this was less of challenge at ten years out than it was at five years out, this is nonetheless a clear example of the kinds of administrative costs that creative producers experience.

REFERENCES

Allison, I. (2017). Major music rights societies join up for blockchain copyright using IBM and Hyperledger. *International Business Times*, 7 April. Retrieved from https://www.ibtimes.co.uk/major-music-rights-societies-join-blockchain-copyrights-using-ibm-hyperledger-1615942.

Allison, I. (2019). Louis Vuitton owner LVMH is launching a blockchain to track luxury goods. *Coin Desk*, 26 March. Retrieved from https://www.coindesk.com/louis-vuitton-owner-lvmh-is-launching-a-blockchain-to-track-luxury-goods.

APRA-AMCOS (2017). Review into the efficacy of the Code of Conduct for Australian copyright collecting societies: APRA-AMCOS response to the discussion paper. Retrieved from http://apraamcos.com.au/media/government/18_Code-of-conduct-review_discussion-paper-response.pdf .

ARIA (2019). ARIA 2018 music industry figures show 12.26% growth. *Aria Charts*, 4 April. Retrieved from https://www.ariacharts.com.au/news/2019/aria-2018-music-industry-figures-show-12-26-growth.

Australian Communications and Media Authority (ACMA) (2019). Communications report 2017–18. Retrieved from https://www.acma.gov.au/sites/default/files/2019-08/Communications%20report%202017-18.pdf.

Australian Film Television and Radio School (2019). The business of creativity: Industry skills survey results 2019. Retrieved from https://www.aftrs.edu.au/about/get-involved/industry-skills-survey/.

Bakhshi, H. and Cunningham, S. (2016). Cultural policy in the time of the creative industries. Retrieved from Nesta website: https://media.nesta.org.uk/documents/cultural_policy_in_the_time_of_the_creative_industries_.pdf.

Bakhshi, H., Hargreaves, I. and Mateos-Garcia, J. (2013). A manifesto for the creative economy. Retrieved from Nesta website: https://media.nesta.org.uk/documents/a-manifesto-for-the-creative-economy-april13.pdf .

Boeuf, B., Darveau, J. and Legoux, R. (2014). Crowdfunding as a new approach for theatre projects. *International Journal of Arts Management*, **16**(3), 33–48.

Brabham, D. (2017). How crowdfunding discourse threatens public arts. *New Media & Society*, **19**(7), 983–99.

Bridgstock, R. (2013). Not a dirty word: Arts entrepreneurship and higher education. *Arts and Humanities in Higher Education*, **12**(2–3), 122–37.

Catalini, C. and Gans, J. (2017). Some simple economics of the blockchain (MIT Sloan Research Paper No. 5191–16). Retrieved from https://ssrn.com/abstract=2874598.

Caves, R. (2002). *Creative industries: Contracts between art and commerce?* Cambridge, MA: Harvard University Press.

CitiGPS (2018). Putting the band back together: Remastering the world of music. Retrieved from https://ir.citi.com/NhxmHW7xb0tkWiqOOG0NuPDM3pVGJpVzXMw7n%2BZg4AfFFX%2BeFqDYNfND%2B0hUxxXA.

ConsenSys (ConsenSysMedia) (2018). Joe Lubin: Nature of the firm, v2.0 keynote from EtherealNY #Blockchain Conference 2018 (Video file), 13 May. Retrieved from https://www.youtube.com/watch?v=SQbcGhnv4jw.

Cooke, C. (2014). PRS confirms Global Repertoire Database "cannot" move forward, pledges to find "alternative ways". *Complete Music Update*, 10 July. Retrieved from https://completemusicupdate.com/article/prs-confirms-global-repertoire-database-cannot-move-forward-pledges-to-find-alternative-ways/.

Copyright Agency (2018). Pilot will test digital codes to stop inauthentic Indigenous art. Retrieved from https://www.copyright.com.au/2018/06/pilot-will-explore -digital-codes-to-stop-inauthentic-indigenous-art/.

Cunningham, S. and Craig, D. (2019). *Social media entertainment: The new intersection of Hollywood and Silicon Valley*. New York, NY: NYU Press.

Department of Communications and the Arts (DOCA) (2018). Consumer survey on online copyright infringement 2018. Retrieved from https://www.communications .gov.au/documents/consumer-survey-online-copyright-infringement-2018-report.

Digital Media Research Centre (DMRC) (2018). The creative economy in Australia: Cultural production, creative services and income. Retrieved from Queensland University of Technology website: https://research.qut.edu.au/dmrc/wp-content/ uploads/sites/5/2018/03/Factsheet-2-Employment-by-sector-V5.pdf.

Dommann, M. (2019). *Authors and apparatus: A media history of copyright* (S. Prybus, trans.). London: Cornell University Press.

Donoughue, P. (2017). Growth in Spotify, Apple Music users boosting revenue for Australian musicians, figures show. *ABC News*, 4 October. Retrieved from https:// www.abc.net.au/news/2017-10-04/streaming-services-boost-revenue-for-australian -musicians/9013974.

Flew, T. (2012). *The creative industries: Culture and policy*. London: Sage.

George, S. and Rheinberger, B. (2017). 94 films: A Commercial analysis. *Screen Australia*, 28 February. Retrieved from https://www.screenaustralia.gov.au/ getmedia/a242af0d-da88-439d-922e-5c247d33a4c1/170606-94-films-analysis.pdf.

Higgs, P. and Lennon, S. (2014). Australian creative employment in 2011: Applying the NESTA Dynamic Mapping definition methodology to Australian classifications. Retrieved from Queensland University of Technology website: https://eprints.qut .edu.au/92726/.

Howard, G. (2018). *Everything in its right place: How blockchain technology will lead to a more transparent music industry*. Beverly Farms, MA: 9GiantSteps.

Howard, G. (2019). Presentation to the MIT Digital Currency Initiative, MIT, 25 September. Boston, MA.

Keat, R. (2000). *Cultural goods and the limits of the market*. London: Palgrave Macmillan.

McCabe, M. (2019). Netflix: "No amount of data can tell you what to commission next". *Campaign Live*, 19 June. Retrieved from https://www.campaignlive.co.uk/ article/netflix-no-amount-data-tell-commission-next/1586932.

McCall, C., Hall, T., Mahon, A., Currell, J., Pitts, S. and Evans, O. (2018). Don't let tech giants bury public service TV. *The Guardian*, 15 October. Retrieved from https://www.theguardian.com/media/2018/oct/15/dont-let-tech-giants-bury-public -service-tv.

McIntyre, H. (2017). Spotify acquires MediaChain. *Forbes*, 27 April. Retrieved from https://www.forbes.com/sites/hughmcintyre/2017/04/27/spotify-has-acquired -blockchain-startup-mediachain/#4875c6f369ee.

Novak, M., Davidson, S. and Potts, J. (2018). The cost of trust: A pilot study. *Journal of British Blockchain Association*, **1**(2).

O'Dair, M. (2019). *Distributed creativity: How blockchain technology will transform the creative economy*. Heidelberg: Springer.

O'Dwyer, R. (2018). Limited edition: Producing artificial scarcity for digital art on the blockchain and its implications for the cultural industries. *Convergence: The International Journal of Research into New Media Technologies*, 1–21.

Pettit, C., Liu, E., Rennie, E., Goldenfein, J. and Glackin, S. (2018). Understanding the disruptive technology ecosystem in Australian urban and housing contexts: A roadmap. Australian Housing and Urban Research Institute.

Rauchs, M., Glidden, A., Gordon, B., Pieters, G., Recanatini, M., Rostand, F., Vagneur, K. and Zhang, B. (2018). Distributed ledger technology systems: A conceptual framework (Cambridge Centre for Alternative Finance report). Cambridge, MA: Cambridge University Judge Business School.

Resnikoff, P. (2016). *Major label CEO confirms that "playlist payola" is a real thing...* Digital Music News, 20 May. Retrieved from https://www.digitalmusicnews.com/ 2016/05/20/playlist-payola-real-killing-artist-careers/.

Rethink Music (2015). Fair music: Transparency and payment flows in the music industry. Berklee Institute of Creative Entrepreneurship. Retrieved from https:// www.berklee.edu/.

Roy Morgan (2019). Netflix surges beyond 11 million users in Australia. Roy Morgan, 19 March. Retrieved from http://www.roymorgan.com/findings/7912-netflix-foxtel -stan-youtube-amazon-february-2019-201903180631.

Screen Australia (2012). Getting down to business: The Producer Offset five years on. Retrieved from https://www.screenaustralia.gov.au/getmedia/14380132-5665-4504 -83c9-799b5b0cba4e/Getting-down-to-business.pdf?ext=.pdf.

Screen Australia (2015). Issues in feature film distribution. Retrieved from https://www .screenaustralia.gov.au/.

Screen Australia (2017). Skin in the game: The Producer Offset 10 years on. Retrieved from https://www.screenaustralia.gov.au/getmedia/cbd7dfc8-50e7-498a-af30 -2db89c6b3f30/Skin-in-the-game-producer-offset.pdf.

Screen Australia (2018). Cinema industry trends. Retrieved from https://www .screenaustralia.gov.au/fact-finders/cinema/industry-trends.

Singleton, M. (2015). This was Sony Music's contract with Spotify. *The Verge*, 19 May. Retrieved from https://www.theverge.com/2015/5/19/8621581/sony-music -spotify-contract.

Spangler, T. (2019). Spotify shuts down ability for independent artists to upload music directly. *Variety*. Retrieved from https://variety.com/2019/digital/news/spotify-shuts -down-artist-direct-upload-1203256886/.

Spotify (2019). Why we're joining the Libra association. Retrieved from https:// newsroom.spotify.com/2019-06-18/why-were-joining-the-libra-association/.

Szabo, N. (1997). Formalizing and securing relationships on public networks. *First Monday*, 2(9). Retrieved from http://firstmonday.org/ojs/index.php/fm/article/view/ 548/469.

Telsyte (2019). Australians turn to multiple subscriptions for entertainment. *Telsyte*, 19 August. Retrieved from https://www.telsyte.com.au/announcements/2018/7/30/svod -feeds-australians-insatiable-appetite-for-streaming-content-ybrdk.

Throsby, D. and Petetskaya, K. (2017). Making art work: An economic study of professional artists in Australia. Retrieved from Australia Council for the Arts website: https://www.australiacouncil.gov.au/workspace/uploads/files/making-art -work-throsby-report-5a05106d0bb69.pdf .

Turner, A. (2018). Australia's $315bn digital opportunity: Data61 CEO talk at D61+Live. CSIRO, 15 October. Retrieved from https://algorithm.data61.csiro.au/ adrians-talk-at-d61-live/.

UNESCO (2018). *Reshaping cultural policies: Global report 2018*. Retrieved from https://en.unesco.org/creativity/global-report-2018.

van Rijn, P. (2018). Collecting societies are struggling to keep up with the influx of millions of lines of data (Blog post). *Music Business Worldwide*, 13 May. Retrieved from https://www.musicbusinessworldwide.com/the-future-of-digital-performance-rights-management/.

Werbach, K. (2018). *The blockchain and the new architecture of trust*. Cambridge, MA: MIT Press.

6. Automated journalism: expendable or supplementary for the future of journalistic work?

Aljosha Karim Schapals[1]

INTRODUCTION

On 25 January 2018, then British Prime Minister Theresa May gave a keynote address at the World Economic Forum in Davos, Switzerland, in which she urged governments, businesses, investors and society at large to embrace the impact of technology and to harness the powers of artificial intelligence (AI). In the speech, May (2018) noted that technology was already "changing the nature of our workplaces and leaving many people with less predictable work patterns". From healthcare to speech recognition and translation, the use of AI has led to transformations in many sectors of the world's economies.

Transferred to the information domain, much ink has since been spilled on the rise of AI in journalism. The computational increasingly manifests itself in ways of verifying information encountered on social media, such as a journalist's sources. Examples include the AI-powered *News Tracer* tool, which helps journalists to assess the credibility of individual tweets (Keohane, 2017). The *SocialSensor* application developed by a major European Union research project works similarly. Based on a number of set algorithms, the software is able to assign credibility scores to journalistic sources and even to detect stories involving significant social media activity in the first place, thus giving it a "nose for news" (Thurman et al., 2016). While such tools may be seen as welcome additions to aid the process of news work, the rise of automated journalism or *robo-journalism* in recent years has proven to be more controversial. The technology allows news organisations to partly automate the process of news writing, requiring little to no human involvement, with the exception of the initial programming (Carlson, 2015).

But if processes of news production can be partly automated, what of the future of journalistic work and the value of journalists for the creative economy? This chapter discusses this question and contributes topical insights into the future of a creative industry characterised by precarious employment patterns. To better contextualise these transformations, this chapter first presents historical perspectives on the rise of the "computational turn" in journalism. Then, the scope of application and the main uses are briefly explained, before moving on to the chapter's central concern: understandably, anxieties revolve around "the replaceability of humans by the machine" (Bucher, 2017, p. 921). If natural language generation (NLG) software can mirror the outputs originally produced by human journalists, does automated journalism render the work produced by journalists expendable? Or should the technology merely be seen as a supplementary toolkit to aid the – often challenging – work of journalists? Needless to say, given its novelty, research into the merging of AI and journalism appears to be a moving target. As such, this chapter concludes with an agenda for future research to help assess what the increasing use of the technology might mean for the future of journalistic work.

HISTORY AND CONTEXT

The first signs of computation affecting the journalism domain date back as far as the 1960s, credited to the rise of computer-assisted reporting in the news industry. However, it was not until the late 2000s that data journalism as a particular subset of computational journalism began to establish itself as a credible field. *The Guardian*'s "Datablog", launched in 2009, is a prominent example. Computation in journalism quickly established itself as an important discursive order by which "journalism and computer science merge together as a new skillset to engage in a big data world" (Ziegler, 2015, p. 28). What followed was the widely publicised expenses scandal featuring United Kingdom Members of Parliament. In an unprecedented move, *The Guardian* published 460 000 pages of expense reports, asking members of the public to sift through them and to flag questionable claims (Coddington, 2015; Flew et al., 2012). In 2016, the Panama Papers published by the German newspaper *Süddeutsche Zeitung* won the Investigation of the Year prize at the prestigious Data Journalism Awards. The newspaper's revelations about secretive dealings of the offshore economy – a collaborative effort by more than 370 journalists across 100 media organisations based in 80 different countries – serves as a prime example of how big data analytics can aid the process of news work. In so doing, computational tools were deployed to unearth data (Thurman, 2019). The process involved 2.6 terabytes of raw data, and high-performing algorithms were deployed to scan the data and spot well-known names and specific links to help journalists draw up lists of known figures and how their assets were protected by tax-evasion schemes (Burgess, 2016).

A still relatively novel sub-genre of computational journalism is automated journalism, now widely discussed in academic scholarship (Thurman,

2019), and seen as leading computational journalism into a "new phase" (van Dalen, 2012). To this date, automated journalism – a process involving the creation of narrative texts not requiring human involvement except for an initial programming effort (Carlson, 2015) – is primarily used for particularly data-intensive journalistic beats such as routine sports coverage or financial news. This degree of automation, however, rests on the availability of clean, structured and reliable data (Haim and Graefe, 2017). This precondition is vital: should the data input turn out to be erroneous, so too will the generated text, that is, the "journalistic" output. Provided the data is correct, however, automatically generated articles will produce fewer errors than human journalists do. Furthermore, these articles can even be personalised to suit the needs of various target audiences: for example, articles can be generated in multiple languages, thereby ensuring a wide reach beyond a purely domestic audience (Haim and Graefe, 2017).

Predictions about automation in the journalism domain, however, have been less positive, a result of suggested claims referring to the potentially expendable role of journalists in the news production process (see Bucher, 2017, p. 921). If processes of news production are to be partly automated, and NLG can mirror the outputs originally produced by human journalists, are such concerns justified? If so, it is no surprise, then, that automated journalism is "often perceived as a threat to the livelihood of classic journalism" (Graefe, 2016, "For Journalists", para. 1). Conversely, more positive predictions include the assumption that, with an increasing roll-out of automation in the newsroom, journalists would in turn be able to devote themselves to in-depth investigations that involve a highly complex skill-set best embodied by human journalists: these include "news judgement, curiosity, and scepticism" (Thurman et al., 2017, p. 1240). According to Clerwall (2014), "an optimistic view would be that automated content will free resources that will allow reporters to focus on more qualified assessments, leaving the descriptive 'recaps' to the software" (p. 527). There is, *inter alia*, a clear divide in what Gynnild (2014) described as "high-tech optimists and practice-focused sceptics" (p. 715), between those who consider automated journalism a threat – a result of routine tasks being taken over by computation – and those who see it as an opportunity, freeing themselves from descriptive tasks and moving on to more qualified assessments instead (Clerwall, 2014).

In recent years, academic research on the phenomenon of computational journalism generally, or automated journalism more specifically, has increased (Haim and Graefe, 2017; Thurman et al., 2017). To this date, however, little is known about the aforementioned divide between those who may feel concerned about an increase in automation and those who value the technology's opportunities. While the technology clearly has its limitations – most obvious in its lack of ability to perform more complex tasks such as establishing causality or contributing to the formation of public opinion (Graefe, 2016) – previous research also suggests that it would bring unprecedented opportunities for the future of investigative reporting (van Dalen, 2012). Such a zero-sum dichotomy of automated journalism as *either* threat *or* opportunity is not reflective of the realities on the ground; as is so often the case, the actual consequences lie somewhere in the middle. For a more detached, neutral and objective assessment to take place, a nuanced analysis of its scope of application is needed.

SCOPE AND USAGE

Automated journalism is on the rise: in 2014, the Associated Press started automating corporate earnings reports; to this date, about 3700 such earnings reports were produced every quarter (Fanta, 2017). In the United States, media organisations such as *Forbes*, *The New York Times*, *The Los Angeles Times* and *ProPublica* are known to experiment with the innovation (Graefe, 2016). In Germany, the local newspaper *Berliner Morgenpost* and the *Handelsblatt* have piloted the software for financial news and stock exchange reports (Dörr, 2016). But although automated journalism is a growing field, the technology itself thus far remains in its infancy: rather than deploying in-house, tailored software, most news organisations rely on external service providers to generate data-intensive, automated content. Such technology is also used in the field of e-commerce to automatically generate product descriptions. German-based companies AX Semantics, Text-On, 2txt NLG, Textomatic and Retresco are leading providers of NLG.

There are apparent benefits in the use of such software: once a data set has been prepared and processed accordingly, the software is able to produce a vast array of articles in a short space of time – in some cases, up to 2000 articles per second (Hammond, 2017). While on the one hand this could lead to lowered production costs and increased profit margins, on the other hand, concerns have been raised that an increase in automation could endanger newsroom jobs characterised by data intensity. Furthermore, algorithms cannot interrogate data or even affirm causality, and are therefore "limited in their ability to observe society and to fulfil journalistic tasks, such as orientation and public opinion formation" (Graefe et al., 2016, p. 6).

Despite these obvious limitations, however, previous studies point to a higher subjective perceived credibility by readers of such content (Graefe et al., 2016). Equally, human-written articles scored higher for readability. Although an overall score indicated that, taken together, human-written articles were rated more favourably, readers were generally unable to clearly distinguish computer-generated from human-written articles. That said, however, "perceived quality does not necessarily relate to objectively measured quality" (Graefe et al., 2016, p. 6). By a similar measure, journalistic texts scored higher on factors such as coherence, clarity and being "pleasant" to read, while readers rated computer-generated texts as more descriptive, informative, trustworthy and objective (Clerwall, 2014). With forecasted increased sophistication of the software itself, there is a case to be made that readers' perceptions of automated journalism, particularly with regard to readability, will improve in the years to come.

As mentioned earlier, with a few notable exceptions, studies on how journalists perceive automated journalism thus far remain rare. Even less common are studies looking at possible tensions between what newsroom staff in managerial positions may deem advantageous versus what rank-and-file journalists may perceive as a genuine threat to the autonomy of their profession. According to Graefe (2016), "due to its ability to produce low-cost content in large quantities in virtually no time, automated journalism appears to some researchers as yet another strategy for news organisations to lower production costs and increase profit margins" (p. 25). Existing research has shown that while some herald the benefits of automated journalism to significantly increase the quantity and speed of news production (Graefe, 2016), journalists themselves so far remain sceptical of the technology (Thurman et al., 2017). This is comprehensible in so far as journalists have in the past voiced concerns over the viability of their profession in the context of technological innovation (van Dalen, 2012). However, a growing body of literature (Carlson, 2015; Graefe, 2016) points to the potential for journalists increasingly to engage with tasks that automated journalism is unable to perform, such as investigative reporting. Naturally, this provides a welcome window of opportunity for the profession to re-discover its central role as a watchdog over society, particularly in the context of the current debate surrounding the issue of widespread misinformation. In light of such prevailing ambiguity of both the benefits and limitations automated journalism brings, one needs to take a closer look at the merits of these seemingly opposing schools of thought.

EXPENDABLE OR SUPPLEMENTARY?

As the journalism domain and the discipline of computer science become ever more closely intertwined, reconsidering and re-examining the journalistic skill-set becomes more vital to make sense of what journalism is becoming in an era of digital transformation. Do editors and journalists, in fact, feel that their autonomy is threatened by profound technological innovation? Or do they regard innovative approaches to the journalism domain as *complementary* rather than *competing* (Neuberger and Nuernbergk, 2010)? Indeed, are concerns that "automated content creation is seen as a serious competition and a threat to the job security of journalists performing basic routine tasks" (van Dalen, 2012, p. 653) valid and veritable? Such routine tasks commonly define the work of rank-and-file journalists and are particularly prevalent in factual accounts of sports coverage, for example. Journalists in this domain, however, were not particularly concerned about an increase of automation in their profession. But such a Manichean view of automated journalism as *either* threat *or* opportunity omits that "new technologies have always been met with overtly optimistic or pessimistic scenarios arguing that the new development will change media content for better or for worse. Automated content creation is no exception" (van Dalen, 2012, p. 654). There are nuances inherent to the adoption of automated journalism that require a more distinct analysis of the two differing schools of thought that the technology seems to have attracted thus far.

One of the few studies to have done so has looked at the implementation of automated journalism across Europe (Fanta, 2017), finding that newsrooms show a general willingness to invest in automation in the future. Elsewhere, it is already firmly established, such as in the case of Norwegian news agency NTB, which has developed an algorithm that helps to generate stories for all the country's 20 000 annual league football games (Fanta, 2017). As such, "journalists can...be put in the position of automating parts of their work they deem unrewarding or repetitive and focus on more high-value tasks" (Fanta, 2017, p. 20). This raises the question of how reporters would be best advised to develop their core journalistic skill-set, as well as the implications this may have for the future of journalism education. According to Dörr (as cited in Fanta, 2017), journalists should make a concerted effort to invest in their personal computational, mathematical and data literacy skills (p. 16). Research by Thurman et al. (2017), however, makes an opposing statement: precisely because algorithms are expected to improve the readability of computer-generated texts in the years to come, journalists would in turn best be advised to focus on skills that are somewhat inherent to the journalism profession and that journalists can best perform. That is, rather than placing emphasis on measurable and quantifiable data literacy skills, journalists should instead strengthen their "news judgement, curiosity, and scepticism"; in other words, invest in the skills "that *human* journalists embody" (Thurman et al., 2017, p. 1240, emphasis added).

Contrary to a notion of "man versus machine", what is central to such reflections is "the need for journalists to concentrate on their own strengths rather than compete on the strengths of automated content creation" (van Dalen, 2012, p. 653). The inability of computational software to perform specific tasks points to the boundaries of automated journalism and explains why, though increasing, the phenomenon remains somewhat of a niche project (Linden, 2017) at this stage of its evolution. Due to the obvious limitations of the software, Thurman et al. (2017) found that journalists remained sceptical, lamenting the lack of creativity in the generated articles. Further limitations identified by Thurman et al. (2017) included the reliance on singular, isolated data streams; the one-dimensional nature of the data; and the need to template stories prior to their automatic generation. In addition, the lack of human angles makes it unlikely that automated systems genuinely "understand all the nuances of human expression" (Thurman et al., 2017, p. 1254).

That said, while the technology clearly has its limitations – most obvious in its lack of ability to perform more complex tasks such as establishing causality or contributing to the formation of public opinion (Graefe, 2016) – some scholars tend to argue that it would bring unprecedented opportunities for the future of investigative reporting (van Dalen, 2012). In an ideal scenario, journalists would be able to devote more time and resources to in-depth analysis and features, while less complex reporting on sports, weather or stock markets would increasingly be performed by algorithms. Such a view would subscribe to the notion that "automated content will free resources that will allow reporters to focus on more qualified assessments, leaving the descriptive 'recaps' to the software" (Clerwall, 2014, p. 527). However, less positive predictions of news automation have instead focused on the increased likelihood of newsroom redundancies as a result of the gradual introduction of automation in the newsroom. Taken together, these schools of thought point to a tension in what constitutes both the challenges and opportunities of automated journalism. Triangulating these through a number of in-depth interviews conducted with German journalists in early 2018 resulted in a comprehensive overview of the status quo of automated journalism at this stage of its evolution (Schapals and Porlezza, 2020). Some retrospective research reflections are presented as follows.

SOME RESEARCH REFLECTIONS

These interviews demonstrated that journalists (a) saw the rise of automated journalism as an *opportunity* rather than a threat, highlighting that (b) the technology may be a useful, *supplementary* toolkit, but that it would by no means render the work of journalists expendable or redundant. As such, contrary to initial expectations, newsroom staff did not feel threatened by the technology – irrespective of whether they worked in a senior or a rank-and-file role within the newsroom. Instead, the interviewed journalists subscribed to the idea that an increase in automation would allow them to spend more time on journalistic investigations that require the skills of human journalists. Indeed, journalists were eager to stress the normative foundations upon which journalism as a *craft* were built. Some examples were that journalism is a creative process; journalism is a uniquely individual craft; and that journalism must include details about background and context so that audiences can contextualise information accordingly (Schapals and Porlezza, 2020).

In future, studies looking at journalism education over time would add a valuable dimension to this debate: how are the skill-sets of journalism graduates changing as a result of the computational turn, if it all? Will such changes be reflected in the curricula of educational institutions? Or will a sharpened focus on the skills embodied by human journalists – "news judgement, curiosity, and scepticism" (Thurman et al., 2017, p. 1240), among others – be more advisable for journalism educators? While it would be impossible to make any such predictions at this stage in its evolution, what is certain is that the technology is likely here to stay. However, given its perceived value as merely supplementary, somewhat inflated man-versus-machine stories overlook the crucial interpretative and sense-making role that journalists continue to hold in society.

NOTE

1.   ORCID iD: 0000-0001-9512-8792. School of Communication, Queensland University of Technology, aljosha.schapals@qut.edu.au.

REFERENCES

Bucher, T. (2017). "Machines don't have instincts": Articulating the computational in journalism. *New Media & Society*, **19**(6), 918–33.

Burgess, M. (2016). How the 11.5 million Panama Papers were analysed. *Wired.* Retrieved from http://www.wired.co.uk/article/panama-papers-data-leak-how -analysed-amount.

Carlson, M. (2015). The robotic reporter: Automated journalism and the redefinition of labour, compositional forms, and journalistic authority. *Digital Journalism*, **3**(3), 416–31.

Clerwall, C. (2014). Enter the robot journalist: Users' perceptions of automated content. *Journalism Practice*, **8**(5), 519–31.

Coddington, M. (2015). Clarifying journalism's quantitative turn: A typology for evaluating data journalism, computational journalism, and computer-assisted reporting. *Digital Journalism*, **3**(3), 331–48.

Dörr, K. (2016). Mapping the field of algorithmic journalism. *Digital Journalism*, **4**(6), 700–722.

Fanta, A. (2017). Putting Europe's robots on the map: Automated journalism in news agencies. Retrieved from Reuters Institute website: https://reutersinstitute.politics .ox.ac.uk/our-research/putting-europes-robots-map-automated-journalism-news -agencies.

Flew, T., Spurgeon, C., Daniels, A. and Swift, A. (2012). The promise of computational journalism. *Journalism Practice*, **6**(2), 157–71.

Graefe, A. (2016). Guide to automated journalism. Retrieved from Tow Centre for Digital Journalism website: https://www.cjr.org/tow_center_reports/guide_to _automated_journalism.php.

Graefe, A., Haim, M., Haarmann, B. and Brosius, H-B. (2016). Readers' perception of computer-generated news: Credibility, expertise, and readability. *Journalism*, **9**(5), 595–610.

Gynnild, A. (2014). Journalism innovation leads to innovation journalism: The impact of computational exploration on changing mindsets. *Journalism*, **15**(6), 713–30.

Haim, M. and Graefe, A. (2017). Automated news: Better than expected? *Digital Journalism*, **5**(8), 1044–59.

Hammond, P. (2017). From computer-assisted to data-driven: Journalism and big data. *Journalism*, **18**(4), 408–24.

Keohane, J. (2017). What news-writing bots mean for the future of journalism. *Wired*. Retrieved from https://www.wired.com/2017/02/robots-wrote-this-story/.

Linden, C-G. (2017). Decades of automation in the newsroom: Why are there still so many jobs in journalism? *Digital Journalism*, **5**(2), 123–40.

May, T. (2018). PM's speech at Davos 2018: 25 January. Retrieved from https://www.gov.uk/government/speeches/pms-speech-at-davos-2018-25-january/.

Neuberger, C. and Nuernbergk, C. (2010). Competition, complementary or integration? The relationship between professional and participatory media. *Journalism Practice*, **4**(3), 319–32.

Schapals, A.K. and Porlezza, C. (2020). Assistance or resistance? Evaluating the intersection of automated journalism and journalistic role conceptions. *Media and Communication*, **8**(3).

Thurman, N. (2019). Computational journalism. In K. Wahl-Jorgensen and T. Hanitzsch (eds), *The handbook of journalism studies* (2nd edn). New York, NY: Routledge.

Thurman, N., Dörr, K. and Kunert, J. (2017). When reporters get hands-on with robo-writing: Professionals consider automated journalism's capabilities and consequences. *Digital Journalism*, **5**(10), 1240–59.

Thurman, N., Schifferes, S., Fletcher, R., Newman, N., Hunt, S. and Schapals, A.K. (2016). Giving computers a nose for news: Exploring the limits of story detection and verification. *Digital Journalism*, **4**(7), 838–48.

van Dalen, A. (2012). The algorithms behind the headlines: How machine-written news redefines the core skills of human journalists. *Journalism Practice*, **6**(5–6), 648–58.

Ziegler, D. (2015). Computational journalism: Shaping the future of news in a big data world. In W.J. Gibbs and J. McKendrick (eds), *Contemporary research methods and data analytics in the news industry*. Hershey, PA: IGI Global.

7. Robotics and artificial intelligence in architecture: what skills will architects need in 2050?[1]

Cori Stewart,[2] Glenda Amayo Caldwell,[3] Müge Belek Fialho Teixeira[4] and Jonathan Roberts[5]

INTRODUCTION

Robots and artificial intelligence (AI) are opening new frontiers in architecture: from accelerating the creation of ground-breaking and iconic buildings, to increasing the potential for the widespread democratisation of design using new digital platforms. Architecture, like many other professional sectors, is undergoing a digital revolution (Susskind and Susskind, 2015), though at a relatively slower or more dispersed rate. The complexity of the architect's job has, to date, made it somewhat resistant to the automation and platform trends that have transformed adjacent industries and professions. Nevertheless, new technologies are changing the skill-set required by architects in surprising ways and ushering forth unprecedented possibilities. New technologies enable previously unimaginable architectural forms, and the re-imagining of places and spaces more suitable for the Anthropocene epoch, the geological era dominated by humans (Lewis and Maslin, 2015).

The new Robot Science Museum, which will be built in Seoul, offers an instructive example of the ways in which technology and robotics are reshaping the field of architecture. Architect and Melike Altınışık Architects founder, Melike Altınışık, states that the museum will not just "exhibit robots, but actually from the design, manufacturing to construction and services[,] robots will be in charge" (Block, 2019, para. 6; Walsh, 2019, para. 5). According to Altınışık the museum "will start its first exhibition with its own construction by robots on site" (Block, 2019, para. 7; Walsh, 2019, para. 5).

As innovations transition out of research labs and into industry, robotic and AI applications and technologies have the possibility of being used at every stage of the architectural design and construction process. At the same time, digital technologies are poised to have a disintermediating effect on the profession of architecture as new "platform" business models, along the lines of Uber, Airbnb or Airtasker, are entering the design and construction marketplace. Susskind and Susskind (2015) refer to WikiHouse (see WikiHouse, n.d.) as an example of an open platform that has been set up with the intention to allow anyone the ability to design, manufacture and construct a beautiful house that responds to their needs at a low cost and without the direct involvement of an architect. The existence of such options is changing how architects engage with clients, construction professionals and other stakeholders.

However, as architectural critic Owen Hopkins (2018) has argued, the new age of architecture will bring a wholesale reassessment, not merely of our concept of architects or the buildings they create, but also of the "cultures and economies that exist through them" (para. 18). That is, an architectural revolution, broader than digital innovation, is in progress, and we must ask: How can architects best equip themselves for it? These changes certainly do not mean the end of architecture. As we hope to show, architectural progress demands a unique combination of skills: digital design skills, and scientific, mathematical and engineering expertise, acquired through extensive formal education and professional training. These skills are necessarily matched by innately human abilities: creativity, cultural knowledge, and originality. This hybridity, we argue, is what makes the profession of architecture both relatively resistant to wholesale technological outsourcing and uniquely positioned to benefit from the innovations the digital age will bring. Rather than making architects obsolete, technological change is likely to underscore and intensify architects' importance as cultural and creative producers. Architects, after all, are people who help us make meaning of our complex environments and relationships, as they bring new buildings and artefacts into being. The focus on architects as cultural and creative workers – who are highly versed in the key future skills of *originality*, *systems analysis*, *critical thinking* and *complex problem-solving* (Bakhshi et al., 2017a) – mitigates immediate concerns about the automation of various aspects of architectural processes.

This chapter opens with an overview of the state of the field today, including a review of the skills, knowledge and abilities currently required of architects. We then examine key technological trends in robotics, AI and software developments to identify some of the innovations currently impacting the practice and profession of architecture, before moving on to review recent developments in the research on the future of work and the future of skills in light of automation trends. We then reflect on the architectural skills for 2030 identified by Bakhshi et al. (2017a) and project further ahead, outlining a list of key skills for architects in 2050.

NEW TECHNOLOGIES IN ARCHITECTURE: FROM 2020 INTO THE FUTURE

Compared to other occupations, architecture has been a relatively techno-logically stable profession. This stability can be traced in part to the fact that architecture is a highly regulated industry: architects are required to be regis-tered with a governing entity and to have the necessary tertiary qualifications and approved industry experience before they can practice professionally. Architects are, in this sense, quite distinct from building designers or other design practitioners. As described by the Architects Accreditation Council Australia (2018): "Architects must be both talented designers and skilled communicators, able to balance client wishes, aesthetic values, planning and environment requirements, building codes, good design principles and con-struction costs in the delivery of a project" (p. 4).

Becoming a registered architect requires many years of academic training and time on the job in order to be eligible to sit the architectural registration exams. The Australian Institute of Architects (n.d.) notes the diversity of skills required[6] to become a successful architect – specifically coordination and interpersonal skills, an attention to detail and the "big picture", the ability to communicate and negotiate effectively and lateral thinking – many of which have little or nothing to do with the romantic vision of the creative genius con-juring fantastically imaginative works of art. Architecture is also increasingly requiring technologically enabled and skilled architecture graduates to manage and direct the use of technology into future practice. These technologies started from basic computer-aided design (CAD) software, which replaced the T-square and drawing boards required to create precise architectural drawings; more recent technologies have included sophisticated parametric software for 3D (three-dimensional) modelling, optimisation software, virtual and augmented realities, open-sourced architecture platforms and the robotic construction of buildings.

Indeed, Susskind and Susskind (2015) argue that architecture is one of the professions that will and should be transformed by technology. They see professions as human constructs and structural systems that hold specific and expert levels of knowledge. With the onset of the internet and other technologies, Susskind and Susskind (2015) argue that these constructs and systems can and should be reconsidered when architectural knowledge is available to all end-users through increasingly capable machines. They refer to a post-professional society in which there will be two streams of impact on the professions: automation and innovation. Automation refers to increasingly intelligent or capable machines that can do more than undertake repetitive tasks and optimise processes (Susskind and Susskind, 2015). Innovation refers to the extension of knowledge across diverse networks and platforms, increasing the sharing of practical expertise across a vast population. With technological evolution, the professions will be incrementally distributed in decades from now to roles, tasks and activities that people will complete or oversee as they work with more capable machines (Susskind and Susskind, 2015). For example, could architects' role shift to designing the systems and evolutionary algorithms that will generate populations of designs, with AI choosing the optimum solution based on the criteria set by the architect? This design meta-level would require current architectural knowledge, by an under-standing of evolutionary systems, and high-level capabilities as creative and cultural producers. These twin roles are what could in future make architecture both relatively resistant to wholesale technological outsourcing and uniquely positioned to benefit from the innovations the digital age will bring. Important transitions from current practice to such new roles are already emerging via the application of robotics and AI in architecture, and these are now considered.

Architectural Design Fabrication

For architects, a tool is just another medium to communicate their ideas with the outside world and, in many cases, architects have crafted their own tools, gadgets and mediums to develop their design intentions. This explorative process on the making of things is deeply rooted in the way architects per-ceive their environments and respond to them. Hence, engaging with novel technologies creates fresh possibilities of expressing ideas and creating unique designs. The use of robots in architectural design fabrication has become more prevalent in recent years, due to robots' ability to forge a direct path from a digital design to fabrication. For example, in the MX3D Bridge project by Amsterdam-based architectural practice MX3D (n.d.), industrial robotic arms are used for 3D printing by the deposition of welded stainless steel. The industrial robotic arms move out over the structure as it is built, welding the layers of stainless steel as directed by the design of the digital model. Welded forms have a lot of flexibility in relation to creating complex geometries and force distribution. In other work led by Prof. Achim Menges (n.d.), the researchers at the Institute for Computational Design and Construction (ICD) collaborate with Prof. Jan Knippers from the Institute of Building Structures and Structural Design at the University of Stuttgart to look into nature to learn optimised solutions for unique aesthetic experiences and efficient construction techniques. Elytra, designed by Menges for the Victoria & Albert Museum in London, is a filament pavilion that is using biomimetic design principles of "less material" and "efficient form". The main components of the structure are constructed in ICD's Robotics Lab in Stuttgart by two industrial robotic arms with custom-made end effectors developed for weaving carbon fibre into rigid structures. This set-up affords modular, lightweight, customised components for fast assembly on the construction site.

The direct connection between design and fabrication, as demonstrated by these examples, creates a fundamental shift in the way the design process is perceived, providing architects the ability to remain in control of the whole process of making. Additionally, industrial robotic arms have the capability to make things architects would not normally be able to make, offering immense possibilities, such as designing and manufacturing for extreme environments, faster pace in delivery and customised construction systems for optimised material use. The making process becomes closer to the sketching process, which is more iterative, fluid and directly connected to the architects' minds. Still at their early stages of exploration, the exemplars above are changing the nature of architectural fabrication and informing the future skill requirements for architects. Not only will architects be expected to design buildings and manage their construction through traditional means, but they will also be expected to control the digital design process through computational programming and the construction by controlling industrial robots. To achieve a balance between these two extremes, architects will still be required to be creative problem solvers who are able to manage teams of people and collaborators.

Robotics and Mass Customisation

In terms of the current work environment in 2020, the potential for robotics and AI to transform architecture and the built environment at scale is dependent on the manufacturing and construction industry's ability to take knowledge from these explorations in order to make and install components at an acceptable price point. While robots have been used in manufacturing for half a century, early robots were limited to routine repetitive tasks, in a capacity that was perfected in the mass manufacture of cars. To create custom architectural works, robots need to be able to "see" and "learn", that is, to use AI. Here, the robot must be able to make decisions and perform tasks on custom and bespoke objects, and to do this in a time- and cost-effective way; in industry, this is referred to as mass customisation.

The term *mass customisation* refers to the ability to manufacture one-off or short-run goods at a comparable cost to mass-manufactured goods (or for which customers are willing to pay a premium) (Kotler, 1989; Spring and Dalrymple, 2000). Mass customisation has recently become a widespread trend in manufacturing, with customers demanding personalised goods, from medicines to potato chips, at a decreased cost. But AI-enabled robotics and AI are not the only technological force transforming global manufacturing. Klaus Schwab (2017), executive chairman of the World Economic Forum, notes in his book *The Fourth Industrial Revolution* that new construction materials like graphene have previously unimaginable attributes: "Overall, they are lighter, stronger, recyclable and adaptive" (p. 17), with the power to significantly disrupt the market. Likewise, the Internet of Things, as it relates to manufacturing, refers to the ability of "machines and products [to] communicate with each other cooperatively as everything is interconnected wirelessly" (Gallagher, 2017, p. 7) and is set to radically change the factories of the future. This transformation has been labelled "Industry 4.0", a reference to the three preceding technological revolutions to remake industry and manufacturing: steam (eighteenth and nineteenth centuries), electricity (nineteenth and twentieth centuries) and the computer chip (late twentieth century). The rapid digital transformation of manufacturing is also driving transformation in adjacent industries, such as construction.

The Relationship Between Architecture and Manufacturing for Construction

The key challenge architecture represents for manufacturing is the gap that exists between design capability, and the manufacture of custom artefacts and building elements required for custom-designed architecture. Stewart and Roberts (2018) describe a pertinent example of this: the "crumpled mirror" staircase at the University of Technology Sydney. When CAD rapidly evolved in the late 1990s, it meant that previously impossible-to-conceive ideas could find a form on the computer screen. But the act of actually *creating* computer-designed works can be more difficult – and costly. Architecture firm Gehry Partners used digital design software to design the staircase for the University of Technology Sydney. But, when creative company UAP manufactured that staircase, it had to employ a thousand-year-old technique of meticulously hand-beating every surface until it matched the shape of the computer model.

Reproducing or re-sizing works of art can also present manufacturing challenges. Originally, artisans would carefully measure the original object and then hand craft the copies, sometimes adjusting the scale. Now, modern scanning technology can create highly accurate computer models of such objects, but the same problem of how to manufacture the new objects presents itself. The technology to take a digital design into a mechanical fabrication process exists, but it is normally too costly for one-off pieces and is reserved for mass production. This is where robots come into play.

Traditionally, robots have been used for manufacturing tasks where the shape of the object being worked on is very well understood. For example, robots can be used to remove the excess metal (a process known as "fettling") after metal casting of car engine blocks. A robot can be programmed to do this because the desired final shape of the engine block is known: without visual information, the robot can move the engine block over a grinder to remove any excess metal. But many of the objects created by artists do not have a detailed computer model for the robot to work from. Also, works of art are typically not uniform or predictable in shape. So any robot working on a piece of art will first need to see it from all angles and accurately discover its shape. The technology to see, or scan, objects exists now. The next horizon could see robots transform an object they see, into one that is desired for its aesthetic qualities. While the advent of robotics is enabling dramatic transformations in the creative and formal possibilities of modern architectural practices, more incremental changes in the day-to-day work practices and skill-sets of future architects will be initiated by technological shifts in the realms of AI, automation and the emergence of new business models.

AI and Architectural Work

Although the definition of AI remains highly contested (Miailhe and Hodes, 2017), in the architectural context, we can think of it most usefully as referring to computer systems capable of engaging in human-like thought processes (e.g., translation, visual perception) and of "learning", or at least imitating, intelligent human behaviour (Kok et al., 2009). As AI capacities evolve, aspects of the architectural profession will become automated. Consider "scan-to-BIM" technology, in which laser scanners create images of building components to be integrated into 3D Building Information Modelling (BIM) software. Previously, this technology handled only large, structural components such as walls and floors. But Adán et al. (2018) outline a new "6D" object-recognition process, completely automated and free of human intervention. This process allows for the representation of smaller mechanical, electrical and plumbing elements (e.g., plugs, switches, ducts and signs) that are crucial to a usable building model and can be positioned and inserted in BIM models. The automation of these routine tasks allows for the energies of human designers to be redirected to non-automatable realms of creativity, collaboration and client interaction.

In other work, MIT researchers have recently developed a digital "smart tool" that can integrate "lifecycle analysis" of a building, which forecasts the likely environmental impacts of a building over its life, into the design process (Hester et al., 2018). Thus, important information can be put *into* the design process, rather than trying to make changes afterwards, when they are more difficult to implement (Hester et al., 2018). The implications for improved sustainability of future buildings are significant. This is just one recent example of how architectural ambitions and practices are, with the help of software, adapting to the ever-increasing imperative of building design in the Anthropocene epoch, when, as Bakhshi et al. (2017a) note, environmental concerns are one of several global structural shifts affecting the future of work. Finally, the rise of BIM software has, in addition, already created an entirely new subset of jobs in the architecture profession: while the software itself takes care of automating certain processes, a skilled workforce of BIM managers, technicians and coordinators is needed to oversee the use of the technology (Uhm et al., 2017).

In the examples above it is unclear whether the capabilities of AI (or a new technology) will focus on automating mundane or repetitive tasks (e.g., placing light switches in a building model), thereby freeing architects to redirect time and resources to the creative processes and requirements of design. AI can source and synthesise zoning data and building codes to support early conceptual design. Parametric design software can create countless possible iterations of a structure within minutes and has the potential to significantly reduce the design time required. Yasser Zarei (2012) questions how parametric design alters the role of the designer/architect, which, in turn, forces us to reassess the very nature of the relationship between architects and their tools. This could lead to new business models, as consumers turn to online platforms that allow end-users to design their own homes and interiors – the rise of the platform ecosystem that has so disrupted other industries has yet to be fully reckoned with in architecture. However, opportunities exist to adopt new models that improve outcomes for designers and end-users without undermining the value of the profession (see Maxwell, 2018). In summary, the impacts of the above technologies on architecture will be significant but they will not make architecture obsolete. While the architect's contribution to culture and society as a creative worker will remain a core aspect of the profession, architects will continue to play an essential role in driving new technology, platforms, materials and tools. Recognising these two conjoined trajectories, we now turn to a consideration of the skills required for future architectural work.

Future of Work? Or Future of Skills?

There have been recent alarmist claims about the irreversible and inevitable decline in jobs. Perhaps, these concerns are due to misunderstanding of the well-publicised study by Frey and Osborne (2013) and one of its central claims, an estimate that nearly 50 per cent of employment in the United States (US) is vulnerable to technological change and computerisation. In Frey and Osborne's (2013) ranking of 702 occupations from most to least computerisable, with #1 being most susceptible, architecture was ranked at #82. More recently, Frey and Osborne (2015) admitted that they may have overestimated likely levels of automation.

Bakhshi et al. (2017a) used a mixed-method approach, combining human expertise (panels of experts collaboratively ranking job sectors predicted to grow, remain stable or shrink) and "machine learning" or algorithms, to arrive at more balanced conclusions regarding the status of a range of occupations in the US and United Kingdom in 2030. In addition, Bakhshi et al. (2017a) sought to consider a range of structural factors – technological and non-technological – on future professions. For example, the "peak globalisation" theory argues that the rapid rise of globalisation has run its course; an ageing population could lower the GDP by 40 per cent globally, even with productivity at current rates (Dobbs et al., 2015); and geopolitical uncertainties spill over into policy uncertainty.

Thus, Bakhshi et al. (2017a) note, because both occupations and environments are increasingly complex, a model for understanding future employment also needs to be complex. With this in mind, their approach focused not just on industries, sectors or even individual occupations, but also on skills. Bakhshi et al. (2017a) reason that, in a future in which automation is not only inevitable but also in fact crucial to improved national and regional productivity, the "future of work" question could more profitably be reframed as that of the "future of skills". This approach is also supported by Susskind and Susskind (2015), arguing that professions will be distributed to tasks and activities.

To this end, Bakhshi et al. (2017a) included in their study 120 skills, abilities and "knowledge features", rather than the nine of Frey and Osborne (2015). The 2017 report therefore provides a far richer picture of the probable skill-set demands for future occupations. Further, in pointing out that the workforce as a social institution has proven incredibly resistant to rapid structural change, even during previous technological revolutions (for instance, from steam to electricity and from electrical to digital), Bakhshi et al. (2017a) argue that such change has historically been incremental rather than acute. This observation is particularly apposite to architecture. This highly regulated industry, with onerous education requirements for entry and advancement, is less vulnerable to wholesale digital disruption (e.g., of the kind presented by Uber to the taxi industry) or supply chain disintermediation in which certain jobs, services or companies are no longer needed (e.g., as occurred within the music industry with the advent of iTunes and streaming services). Instead, digital transformation in architecture is likely to occur in a manner not dissimilar to other eras of technology transformation: that is, through generational and educational evolution.

PREDICTING THE FUTURE OF SKILLS FOR ARCHITECTS

In this findings section, we use Bakhshi et al.'s (2017a) findings for US architects in 2030 and project forward to 2050 to assess the likely future skills of architects based on the analysis in the chapter so far (see Table 7.1). While the skills in each column are the same, we argue that their order of importance will change, and we have provided brief justifications for these shifts.

Table 7.1   *Top 20 skills in 2030 and a prediction of skills for architects in 2050*

Architecture and engineering occupations top skills US in 2030 (Bakhshi et al., 2017a)	Proposed top skills for architects in 2050 (authors' predictions)
Information ordering	Originality
Systems evaluation	Active learning
Critical thinking	Critical thinking
Complex problem-solving	Service orientation
Persuasion	Negotiation
Physics	Persuasion
Science	Active listening
Fluency of ideas	Management of personnel resources
Service orientation	Communications and media
Number facility	Judgement and decision-making
Systems analysis	Inductive reasoning
Deductive reasoning	Deductive reasoning
Active learning	Systems evaluation
Mathematical skills	Administration and management
Negotiation	Monitoring
Inductive reasoning	Number facility
Category flexibility	Science/physics
Operations analysis	Information ordering
Mathematical reasoning	Category flexibility
Originality	Fluency of ideas

Source: Based on concepts from Bakhshi et al. (2017a).

Originality

This skill has moved to the first position because we see it as a key factor differentiating humans from machines. This skill is defined as: "The ability to come up with unusual or clever ideas about a given topic or situation, or to develop creative ways to solve a problem" (Bakhshi et al., 2017b, p. 5). Although AI is expected to give machines more capabilities in processing huge sets of information, creativity is a human ability that, due to its unexpected or unusual nature, is difficult to predict and replace.

Active learning
This skill has also moved up the list. As technology continues to evolve rapidly, architects will also need to rapidly adopt new technologies, modes and processes into their work practices. Architects will have to actively seek out learning new skills to bring into their evolving career paths.

Service orientation
We propose that this skill will also become more of a priority. With the advent of platforms that enable and provide more agency to clients in the design process, the role of the architect will shift from being the designer and controlling the design process to more of a consultant who helps and advises clients on the design.

Negotiation and persuasion
These skills have also moved up the list. We see these human skills as becoming more important as automation increases and architects will be working across diverse platforms, machines, robots, data sets, software, clients and stakeholders. Architects will need to negotiate across different capabilities to persuasively address diverse requirements and needs of their clients and collaborators.

Communication and media
We expect that as more technical and labouring work will be outsourced, a larger proportion of human attention will be spent on communication. Architects will need to better explain and communicate the rationale and value of their design and construction decisions. These skills will also be increasingly critical considering that architectural work will continue to grow and the environmental and climate change agenda will also increase. We anticipate that governments and the general public will be more interested in the sustainability credentials of new buildings. Architects will have to explain in layperson's language how their designs meet these parameters and sustainability requirements.

Judgement and decision-making
These skills will remain essential for an architect. AI will allow for the production of a vast number of design variations, optimisations and formal choices, so the work of the architect will be to narrow them down using their best judgement to make the best choices in terms of costs and benefits.

Inductive and deductive reasoning
Inductive reasoning is defined as the "ability to combine pieces of information to form general rules or conclusions (including finding a relationship

among seemingly unrelated events)" (Bakhshi et al., 2017b, p. 4). Deductive reasoning is defined as the "ability to apply general rules to specific problems to produce answers that make sense" (Bakhshi et al., 2017b, p. 3). These reasoning abilities will continue to inform judgement and decision-making; thus, they will remain important skills for architects because the decisions they will make are likely to become increasingly complex. Although AI and technology may help to manage the number of options and variables, the human and technological factors for consideration, in terms of global challenges, are likely to be more complicated.

Number facility/science/physics

These will continue to be foundational skills of architects. Having a solid grounding in maths and science is vital for making structural decisions, but it will also be increasingly important as architects need to understand, program and control AI and machines.

Fluency of ideas

This skill has moved down the list. This skill emphasises the ability to produce a large number of ideas that are not necessarily all good ideas. Parametric and generative software is currently used by architects to produce different options and variations of designs. Such software will continue to be combined with hardware to rapidly create, test and explore design ideas at increasing rates.

CONCLUSION: BALANCING HUMAN INTERESTS WITH ADVANCED TECHNOLOGICAL SOLUTIONS

When looking at the literature to explore the discourse on skills for architects in the future, we admit that there are fewer academic publications on the topic than initially expected. An explanation for this could be that most of the future-thinking and architecture-related literature focuses on how to design for the future of cities and population growth or the future of architectural education. When searching the internet for the "Future of Architecture", countless online articles and blogposts are found from a range of non-academic sources, including *Fast Company*, *Huffington Post*, *ArchDaily* and *Dezeen*. Other professional sources include the Royal Institute of British Architects' (2018) recent report on architecture and digital transformation, created in collaboration with Microsoft. Although these sources provide insight into current thinking and trends on the future of the profession, there are limited resources that focus on the individual skills required for architects for the future.

Given recent developments with architectural design and fabrication processes, as shown with the exemplar projects discussed previously, one would initially imagine that such skills relevant for the future of architecture would be more focused on computational design thinking, materials research, transdis-ciplinary approaches and sustainable practices. All of these aspects are highly relevant for increasing technological impact on the profession; however, as indicated by the Bakhshi et al. (2017a) report and the skills we propose for 2050, the future of the profession will rely on skill-sets that are uniquely combined by individual architects. For 2050, we expect that human expression is likely to increase, but its translation to the built form will be increasingly automated using technologies such as robotics and AI. We anticipate that the platforms that architects work on will be increasingly sophisticated and increasingly accessible by everyday users.

It is clear that the profession will be faced with ongoing challenges that are not limited to technological changes, but will involve a series of social, political, economic and environmental consequences and drivers. Like others, we can imagine future cities and the roles that architects will play in designing and creating them; however, our purpose in writing this chapter is to question the human skills required to address the wide range of ongoing and increasing challenges that await the profession. Ratti and Claudel (2015) position their concept of *Futurecraft* as situations in which the designers do not predict the future but explore, debate and provocate what the future may entail: "Methodologically, this dissolves prediction-anxiety and opens up the possi-bility of exploring new avenues of research" (p. 29). Like these authors, we are concerned with pursuing this line of research and continuing the discussion necessary to balance human interests, motivations and emotions with AI, robotic evolution and advanced technological solutions.

NOTES

1. This chapter uses an excerpt from Stewart and Roberts (2018). Science makes art. But could art save the Australian manufacturing industry? *The Conversation.* Licensed under Creative Commons 4.0 https://creativecommons.org/licenses/by -nd/4.0/.
2. ORCID iD: 0000-0003-2604-1305, Creative Industries Faculty, Queensland University of Technology, cori.stewart@qut.edu.au.
3. ORCID iD: 0000-0003-0837-9310.
4. ORCID iD: 0000-0003-0593-9597.
5. ORCID iD: 0000-0003-2318-3623.
6. The Australian Institute of Architects website also notes that many of those trained in architecture leave the field to pursue alternative careers in adjacent areas such as asset management, project management, urban planning/design and construction management.

REFERENCES

Adán, A., Quintana, B., Prieto, S.A. and Bosché, F. (2018). Scan-to-BIM for "secondary" building components. *Advanced Engineering Informatics*, **37**, 119–38.

Architects Accreditation Council Australia (2018). Industry Profile: The profession of architecture in Australia. Retrieved from https://www.aaca.org.au/wp-content/uploads/Industry-Profile.pdf.

Australian Institute of Architects (n.d.). Pathways to architecture. Retrieved from https://www.architecture.com.au/explore/pathways-to-architecture.

Bakhshi, H., Downing, J., Osborne, M. and Schneider, P. (2017a). *The future of skills: Employment in 2030*. London: Pearson and Nesta.

Bakhshi, H., Downing, J., Osborne, M. and Schneider, P. (2017b). *The future of skills: Employment in 2030. Glossary of skills*. Retrieved from https://futureskills.pearson.com/research/assets/pdfs/glossary-of-skills.pdf.

Block, I. (2019). Robot Science Museum in Seoul will be built by robots and drones. *Dezeen*, 20 February. Retrieved from https://www.dezeen.com/2019/02/20/robot-science-museum-melike-altinisik-architects-maa-seoul.

Dobbs, R., Remes, J. and Woetzel, J. (2015). Where to look for global growth. *McKinsey Quarterly*, January. Retrieved from https://www.mckinsey.com/featured-insights/employment-and-growth/where-to-look-for-global-growth.

Frey, C.B. and Osborne, M. (2013). The future of employment: How susceptible are jobs to computerisation? Oxford Martin School Working Paper, 17 September. Retrieved from https://www.oxfordmartin.ox.ac.uk/downloads/academic/future-of-employment.pdf.

Frey, C.B. and Osborne, M. (2015). Technology at work: The future of innovation and employment (Citi GPS: Global Perspectives & Solutions). Retrieved from https://www.oxfordmartin.ox.ac.uk/downloads/reports/Citi_GPS_Technology_Work.pdf.

Gallagher, S. (2017). Industry 4.0 Testlabs in Australia: Preparing for the future. Retrieved from http://hdl.voced.edu.au/10707/465887.

Hester, J., Gregory, J., Ulm, F.J. and Kirchain, R. (2018). Building design-space exploration through quasi-optimization of life cycle impacts and costs. *Building and Environment*, **144**, 34–44.

Hopkins, O. (2018). Post-digital architecture will be rough, provisional and crafted by robots. *Dezeen*, 12 December. Retrieved from https://www.dezeen.com/2018/12/12/post-digital-architecture-owen-hopkins-opinion/.

Kok, J.N., Boers, E.J., Kosters, W.A., Van der Putten, P. and Poel, M. (2009). Artificial intelligence: Definition, trends, techniques, and cases. In *Encyclopedia of Life Support Systems*. Paris: EOLSS Publishers.

Kotler, P. (1989). From mass marketing to mass customization. *Planning Review*, **17**(5), 10–47.

Lewis, S.L. and Maslin, M.A. (2015). Defining the anthropocene. *Nature*, **519**(7542), 171.

Maxwell, D.W. (2018). The ecosystem revolution: Co-ordinating construction by design. In D.W. Maxwell (ed.), *Proceedings of the 1st Annual Design Research Conference (ADR18)* (pp. 251–62). Sydney: University of Sydney.

Menges, A. (n.d.). Elytra Filament Pavilion, Victoria & Albert Museum. Retrieved from http://www.achimmenges.net/?p=5922.

Miailhe, N. and Hodes, C. (2017). The third age of artificial intelligence. *Field Actions Science Reports: The journal of field actions* (Special Issue 17), 6–11.

MX3D (n.d.). MX3D Bridge. Retrieved from https://mx3d.com/projects/mx3d-bridge.

Ratti, C. and Claudel, M. (2015). Futurecraft: Tomorrow by design. *TECHNE: Journal of Technology for Architecture and Environment*, **10**, 28–33.

Royal Institute of British Architects (2018). *Digital transformation in architecture*. Retrieved from https://www.architecture.com/-/media/gathercontent/digital-transformation-in-architecture/additional-documents/microsoftribadigitaltransformationreportfinal180629pdf.pdf.

Schwab, K. (2017). *The fourth industrial revolution*. Geneva: World Economic Forum.

Spring, M. and Dalrymple, J. (2000). Product customisation and manufacturing strategy. *International Journal of Operations & Production Management*, **20**(4), 441–67.

Stewart, C. and Roberts, J. (2018). Science makes art. But could art save the Australian manufacturing industry? *The Conversation*. Retrieved from https://theconversation.com/science-makes-art-but-could-art-save-the-australian-manufacturing-industry-97849.

Susskind, R.E. and Susskind, D. (2015). *The future of the professions: How technology will transform the work of human experts*. New York, NY: Oxford University Press.

Uhm, M., Lee, G. and Jeon, B. (2017). An analysis of BIM jobs and competencies based on the use of terms in the industry. *Automation in Construction*, **81**, 67–98.

Walsh, N. (2019). Robots will construct Melike Altınışık Robot Museum in Seoul. *ArchDaily*, 19 February. Retrieved from https://www.archdaily.com/911761/robots-will-construct-melike-altinisik-robot-museum-in-seoul.

WikiHouse (n.d.). WikiHouse about. Retrieved from https://www.wikihouse.cc/About.

Zarei, Y. (2012). *The challenges of parametric design in architecture today: Mapping the design practice* (Doctoral dissertation, University of Manchester, United Kingdom). Retrieved from https://www.research.manchester.ac.uk/portal/files/54523431/FULL_TEXT.PDF.

8. Museum curation in the digital age

Rui Oliveira Lopes[1]

INTRODUCTION

The role of the professional curator progresses as much as the environment in which the curatorial work takes place: from the hoarding of objects judiciously selected to furnish funerary chambers in the ancient world, to the formation of collections of curiosities and rare commodities, methodically gathered by wealthy and learned collectors in the seventeenth century (Ambrose and Paine, 2006), and the virtual and interactive exhibitions created in online platforms (Patel et al., 2003; Sabharwal, 2015; Walczak et al., 2006). The curator is still known as a selector and an interpreter of objects and works of art, as well as a mediator to communicate and establish conceptual or intellectual relations and to engage in a dialogue between the works of art and the audience (George, 2015).

With a closer look at the history of exhibiting collections and the responsibilities of those who act as "keepers" of collections, one would conclude that the *modus operandi* of the curator has been changing as a response to a broader socio-cultural change in the ways the public interacts with the collections (Boylan, 2008). Throughout the nineteenth and twentieth centuries, artists increasingly explored new media and technology, leading to an alluring redefinition of the work of art, the emergence of new forms of interpretation, and new exhibition discourses in which the audience is often and increasingly participative (Henning, 2008).

In this scenario, the curator has proceeded from a figure of intellectual authority, and a custodian of cultural and artistic objects safely kept behind the glass case in a museum, to a cultural mediator who seeks new forms to engage the audience with artists and objects through museum displays and exhibitions. The figure of the subject specialist, the museum collection-based curator, unfolded into a wide range of curator modalities such as the artist-curator (Jeffery, 2015), the independent curator, and even content creators who produce thematic educational and interactive experience exhibitions (O'Neil, 2007). There are not only new modalities of non-institutional curators, but also innovative approaches to communicating and interpreting museum exhibitions in which the intervention of the curator is often deliberately in the shadow. The role of the museum curator may not be authoritative any more, but rather, selective, subliminal and thought-provoking, allowing the visitor to construct meaning based on their experience and knowledge. Through the use of digital technologies, museum crowd-curation enables the public to take the role of the curator and participate in the curatorial process, helping to determine and select the objects to display in the physical museum gallery (Bernstein, 2014; Cann, 2012; Czegledy, 2012).

NEW ROLES OF THE MUSEUM AND THE CURATOR IN THE DIGITAL AGE

The emergence of a digital era introduced significant changes in the roles of the museum and the curator. The museum is increasingly engaging with society and its audience through educational services and the use of interactive media. Museum professionals are increasingly integrating digital technologies into museum practice in several ways, allowing the audience to interact with the museum collections in the physical space of the museum, as well as within the virtual space of the museum.

The museum curator is no longer the central figure in exhibition production, although the curator still carries a pre-eminent responsibility in the definition of exhibition narratives and the mediation between the artists, the collections and the audience. Currently, there is an escalating diversity of jobs linked with museum exhibition production due to the increasing recognition that museums now have a more social, inclusive and educational responsibility within the local and global communities (Boylan, 2008; Coffee, 2008). Many of the tasks assigned to the traditional scholar-curator, such as publications, educational programmes, cataloguing and guided visits, are gradually given to specialist departments. As an example, the Museum of Modern Art (MoMA) in New York currently has six departments: curatorial affairs; education and research support; exhibitions and collections support; external affairs; administrative departments; and retail. The role of the curator concentrates mostly in the curatorial affairs of their specialisation, which involves collecting, managing and exhibiting objects. However, the role of the museum has developed into an expanded form of curating collections. This also includes:

- educational programs to enable interaction between the collections and the audience;
- the provision of research resources such as libraries and museum archives to support the research needs of staff and the general public;

- exhibition designers who are specialised in visual communication to enhance the visitor experience and materialise the curator/artist conceptual framework;
- marketing managers who typically oversee audience development, merchandise and fundraising, and implement other "charming" strategies to increase the attractiveness of both public and potential benefactors.

In this expanded field of the museum occupation, the curator may curate the collections and exhibitions, but all museum staff curate the museum. As Boylan (2008) pointed out, museum employment has been exponentially increasing since the 1960s as a result of a rising number of museums opening and the diversification and specialisation of museum jobs. However, in the United Kingdom (UK), the number of curators in the traditional sense has fallen from around 30 per cent to under 12 per cent over the past 25 to 40 years (Boylan, 2008).

Along with the growth of professionals and miscellany of occupations, museums are increasingly outsourcing, contracting and tendering to deal with financial constrictions or to fill the gaps related to technological groundwork or the specialised workforce (Ball and Earl, 2002; Harrison, 2000). Among the work mostly outsourced by museums are the services provided by designers (exhibitions, branding, publications, leaflets and brochures) and information technology services for computer system management, website development, mobile app development, and gradually, screen productions, digital media and interactive content development (virtual reality (VR), augmented reality (AR), interactive screen productions and kinetic solutions) (Ball et al., 2002). In response to this change, degree programmes are adapting to new challenges, not only focusing on digital literacy and the integration of digital technologies with the physical and digital space of the museum, but also using active collaboration with other professionals related with creative work (Giannini and Bowen, 2015). In 2015, the Pratt Institute launched a new Master of Science in Museums and Digital Culture designed to offer an innovative approach to museum studies that focuses on the digital life of museums across collections, galleries and activities.

Taking the new roles of the museum and the expanded notion of museum curation, this chapter discusses the occupation of curating the museum in the post-internet age by examining the role of creative work in the use of digital media in the museum space, the interaction between the museum and the public through the web and social media engagement, and the curation of digital art. Through interviews with museum professionals, we ascertain their experience and views about the use of digital technologies.

INTERACTING WITH THE DIGITAL IN THE MUSEUM SPACE

In the twenty-first century, under the framework of new museology, visitor engagement and learning through outreach, inclusion, participation and inter-activity are dominant concepts in professional museum practice (Vergo, 1989). In human-centred museum practice, the museum visitor is changing the museum space and the way curators and museum professionals curate the museum through an integrated experience. The visitor experience in the museum space begins in the foyer and continues through all other spaces, from the galleries to the education service rooms, cafeteria, shops and auditoriums (Laursen et al., 2016; Schorch, 2013). In our view, museum curation involves a holistic interpretative interplay between visitors, museological space, content and curatorial discourses that all combine to allow visitors to construct meaning. Making meaning and visitor experience is an expanding critical field in museum studies, leading to the exploration of sensorial forms of interaction between the museum space and the collections.

Digital technologies are commonly pointed to as media designed to offer an interactive educational approach to traditional passive content in the museum space and to heighten the learning, visualisation and perception of objects on display. Digital interfaces empower the museum visitor and enable the selec-tion, use and transformation of visual and textual information, particularly in art museums. Another factor that contributes to the increasing number of digital interfaces in the museums is the interrelationship between museum visitors and funding. As pointed out by Tony Bennett (1995) in *The Birth of the Museum*: "In order to attract sufficient visitors to justify continuing public funding, [museums] thus now often seek to imitate rather than distinguish themselves from places of popular assembly" (p. 104).

Scholars argue that the digital age develops a participatory culture among the public and contributes to building inclusion and plurality in the museum (Bautista, 2014). In this framework, the museum of the future is an institution that is no longer working *for* the community, but *with* the community. Digital communication particularly connects with the younger generation – which is acquainted with digital media and is sensitive to visual stimulation – and is often an important channel to mediatise the museum through personal social media (Jenkins, 2006).

The integration of digital technologies in the physical museum space offers new modes of visualisation, engagement, interpretation and meaning-making (Falk and Dierking, 2008). Over the last few decades, museums have been exploring new ways to introduce digital tools in curatorial practices to create stronger interactions and participation between the museum and the visitors

(Samis, 2008; Walker, 2008). Recent research demonstrates that the museum audience increasingly wants to interact and participate in museum activities in a two-way relationship involving experience and learning (Kelly, 2007; Kelly and Russo, 2008). The visitor experience is amplified by digital technologies involving AR, VR and mixed reality, digital media displays and screen media content using interactive digitised collections and documental video streaming.

The exhibition *Pink Floyd: Their Mortal Remains* at the Victoria and Albert Museum (V&A) (13 May to 15 October 2017) offered a series of digital and interactive interfaces to enhance the visitor experience. The exhibition was advertised as "a spectacular and unparalleled audio-visual journey through Pink Floyd's unique and extraordinary worlds, chronicling the music, design and staging of the band, from their debut in the 1960s through to the present day". The exhibition featured the band's musical instruments, photographs, documents, audio and visual recordings, and a final gallery of surround-sound and images, where visitors could sit down or recline to experience panoramic videos of *The Wall* and 3D (three-dimensional) audio.

The exhibition was arranged as a chronological journey through the history of the band, but it was not just an ordinary biographical narrative. The curator/producer Paula Stainton and the V&A team, led by senior curator Victoria Broackes, gave full detail to Pink Floyd's visual identity, collaborating with the same designers who have worked with Pink Floyd to design their album covers and stage set designs. The curatorial team worked together with Stufish Entertainment Architects, founded by the late Mark Fisher, who designed Pink Floyd's "The Wall Tour" and "The Division Bell Tour" in 1980 and 1994, respectively. Stufish was responsible for designing the exhibition, based on the rich audio and visual history, to bring the exhibition visitor to the world of Pink Floyd, which included the construction of sculptural works and structures such as the Battersea Power Station, the inflatable pigs and the Bedford Van the band used for the tours. The designers included several screens and projectors to display documental videos, photos and audio materials.

The designer and creative director of the exhibition, Aubrey "Po" Powell, was himself the founder of Hipgnosis Studio, which became known for designing the album covers for Pink Floyd, Led Zeppelin, Paul McCartney, Peter Gabriel and many others. Powell designed Pink Floyd's emblematic album covers for *The Dark Side of the Moon*, *Wish you Were Here*, *Animals*, *The Wall*, *The Final Cut* and *The Division Bell*, among others. As a creative director, and having a lifelong creative experience with the band, Powell became a crucial adviser to help Stufish design the exhibition.

The exhibition presented a strong chronological approach to the history of the band, but it was also designed to celebrate the visual icons of Pink Floyd, namely through an interactive installation of the cultural icon that the album cover for *The Dark Side of the Moon* became. The museum tendered Cinimod

Studio to reinterpret the rainbow-emitting prism from *The Dark Side of the Moon* album cover design, using holography. In this production, the specialist hologram team at Cinimod Studio collaborated with Media Power House for digital and technical production. The workforce at Cinimod Studio and Media Power House specialise in architecture and digital art, and include audio-visual engineers, sound engineers, computer scientists, software developers, video technicians and production staff, lighting designers, and media and installation artists.

Besides the interactive visual interfaces, the curatorial team of the exhibition inevitably included audio digital systems specially designed and produced by Sennheiser, one of the most critical partners in the production of the exhibition. Sennheiser created a 3D digital audio upmix of a classic recording and an interactive soundtrack to the exhibition. Visitors were given headphones, and as visitors walked around the exhibition space, the headphones synced to the videos playing in the digital screens.

Sennheiser sound engineers Andy Jackson and Simon Rhodes used AMBEO 3D Audio Technology and a 360-degree surround mix at the legendary Abbey Road Studios to rebuild the immersive audio that Pink Floyd attempted to develop throughout their career. This digital recording was the sensorial element that elevated the visitor experience and made them feel that they were in the middle of the audience during the last performance of "Comfortably Numb" from Live 8 in 2005. Audio and video content along the exhibition space were housed on media players near the screens. The audio was sent over a balanced audio pair to Sennheiser guidePORT transmitters in each gallery's audio-visual rack location. These then sent a radio frequency signal to antennas in the ceiling and other locations in the gallery space. Small trigger units placed near the exhibits, known as identifiers, told the bodypacks which audio stream to receive or play.

The feeling of immersive audio was increased by video projected along the four walls of the gallery using 10 Panasonic 6200-lumen laser projectors, edge-blended, which beamed a continuous band of video 45.5 m long and 3.5 m high. The list of external collaborators in the production of the exhibition continues to grow, with Media Power House supplying all the video equipment and the video installation designed by Richard Turner of Lucky Frog. Richard Turner is a specialist in multi-screen installations using slide projections and computer control systems designed for live performances.

Since the inauguration at the V&A in May 2017, the exhibition has toured in Rome, Dortmund and Madrid. The V&A exhibition attracted more than 350 000 visitors, becoming one of the most popular exhibitions of that museum. It is worth noting that the V&A registered a decline of 12 per cent in visitor numbers in 2016/2017 compared to 2015/2016. The number of visitors to the V&A between 2008/2009 and 2017/2018 demonstrate that the Pink

Floyd exhibition contributed significantly to the highest number of visitors over the last 10 years. The adult price of the tickets for the Pink Floyd exhibition ranged from £22 (weekdays) to £26 (weekends). The price increased towards the last days of the exhibition, up to £30 during the last weekend. The exhibition generated a gross income of approximately £7 million just from the tickets.

This was not the first incursion of the V&A into the trend of blockbuster exhibitions about pop culture icons. Four years previously, the V&A also collaborated with Sennheiser and several other digital industry partners to create interactive media content for the *David Bowie is* exhibition, visited by more than 2 million people over five years at 12 venues around the world.

Blockbuster exhibitions traditionally involve collaborations between museums and creators of audio and visual digital content to offer immersive and interactive experiences to museum goers. In 2017, Tate Modern partnered with HTC Vive to integrate a VR experience in the *Modigliani* exhibition (Giannini and Bowen, 2019). The experience offered a re-creation of the artist's studio using the actual studio space as a template.

The VR experience in the *Modigliani* exhibition involved the collaboration between multiple teams at Tate, including audio-visual, conservation, curatorial and digital content, working together with several professionals from Preloaded, from the creative director to 3D modellers using HTC Vive hardware.

Blockbuster exhibitions attract large numbers of visitors, generating high revenues from box-office and merchandising, which helps to cover the high production costs (Falk and Dierking, 2018; Lawrenson and O'Reilly, 2018). The mediatisation of these exhibitions and positive response from the audience to the integration of interactive digital content contributes to creating a social media buzz, which also contributes to expanding the audience of the museums.

At an increasing pace, museums are evolving from stand-alone VR and screen media units to establish their digital content departments and transform large sections of the museum space to offer visitors an interactive, engaging and educational experience. The *Lumin AR Tour* at the Detroit Institute of Arts, and the National Museum of Singapore's video animated installation *Story of the Forest*, are just some of the examples in which the digital is shaping the work of curators in the museum. Both museums collaborated with GuidiGO to create digital guided tours on mobile devices using AR, image recognition and interactive games to generate engaging experiences.

The extensive use of the internet, mobile devices and the technology currently available enables a vast array of possibilities attracting museum visitors and providing educational experiences. As a result, museum curation has changed not only in the museum's physical space but also in the museum's digital space.

CURATING THE MUSEUM IN THE WORLD WIDE WEB

Technology is currently recognised as one of the most efficient vehicles for making museums known and relevant to the world. It is through the use of digital technologies, mainly through the World Wide Web, that museums reach a global audience. Nowadays, most museums have created their own websites and frequently use other internet channels in social media to communicate and inform the public about their collections, exhibitions, educational services and activities (Bowen, 1997; 2000).

Over the last two decades, museums have strived to digitise their collections and provide online open-access digital content through their websites, mobile apps, social media or 360-degree virtual tours made available at Google Arts & Culture to strengthen visitor interest in the collections, exhibitions and other events in the physical museum. The creation of digital content to supply all these information technology platforms resulted in the need to establish digital content departments and employ professionals with skills in information communication technologies (ICT) and design, at least in some large national museums. In addition to digital media professionals, large national museums frequently hire social media managers, who are given the task to curate the digital content on social media platforms based on big data, user-generated content, user experience and analytics. With the move to digital collections, an increasing number of professionals, within and outside of the museum, work in several ways to create public accessibility for the interpretation and meaning-making of collections. This happens through the curation of digital content involving the construction of narratives, availability of high-resolution digital images, VR, AR, 3D modelling, 360-degree virtual tours and virtual exhibitions. From a museum visitor perspective, the digital content on museum websites, mobile apps or social media platforms is now as relevant as the physical collections and museum space because it contributes to giving visitors a real sense of ownership of cultural heritage (Dallas, 2004; Patel et al., 2003; Walczak et al., 2006; Wojciechowski et al., 2004).

The Metropolitan Museum of Art (The Met) is one of the museums creating more digital content for its website. This includes curatorial highlights, new acquisitions (*MetCollects*), *The Met 360 Project*, which offers virtual visitors an immersive tour through some of the most iconic spaces of the museum, and #MetKids, a digital space of museum collections designed for children aged 7–12. The museum's decisions contribute to a stronger engagement with and interaction between visitors and the collections. For example, it made more than 400 000 high-resolution images of public domain artworks available under Creative Commons Zero, created a digital repository of more than 1500 books at *MetPublications*, and gave visitors the possibility to create their digital collections through *MyMet*. The Met collaborates with Wikimedia, Google, MIT and many other ICT institutions, organisations and corporations to facilitate the dissemination of the museum's collections, build audiences, enable new acquisitions, and eventually, attract patrons and donors. A 2018 press release announced that The Met set a new record with more than 7.35 million visitors during the 2018 fiscal year, while the website (metmuseum. org) registered a total of 30.4 million visits, 32 per cent of which were international. The "online collection" and "Heilbrunn Timeline of Art History" are the most popular pages, with 8.4 million and 9.4 million visits respectively.

The Met Digital Department was established in 2009 out of existing museum staff that had been part of Education, Web Group, Information Systems and Technology, and Collections Management. Loic Tallon (2017), Chief Digital Officer at The Met, argues that "with digital technologies impacting a growing range of verticals within the organization, departments across the organization are required to adapt to digital, and not just a single department" (para. 10). This means that, in a digital age, the museum is not merely required to have a digital department, but, more importantly, the institution itself has to adapt and use the digital effectively.

Digital departments across many cultural institutions have several digital-related tasks and responsibilities requiring active interdepartmental collaboration. In the museum, despite the centralisation of digital-related functions, many other departments, including the curatorial teams, contribute enormously to the creation of digital content. The work of digital departments in the museum is conventionally divided into three groups: collections information management; multimedia content production; and design/development of digital platforms.

The Met advertises job positions through a LinkedIn page. A job advertisement published on 7 November 2019 aimed to recruit a Chief Digital Officer. The statement of responsibilities and duties highlighted that the Chief Digital Officer reports directly to the Director and Deputy Director of the museum. He/she needs to "serve as a change agent in leading a cultural transformation that supports the transition to digital storytelling and technologies, and embraces a digital mindset". The advertisement also stated that the Chief Digital Officer "will hold end-to-end responsibility for all of the Met's digital strategy, content, and product development, to create holistic and intentional digital experiences across channels and in-gallery". The Chief Digital Officer:

> will direct and manage the electronic production of documentation and interpretative material on the collections; facilitate efficient delivery of this information to a variety of audiences, both in the networked environment and in the galleries; and oversee information and collections management, application of standards and best practices, managing electronic documentation and digital assets associated with the collections, and promoting…digital initiatives.

The description of responsibilities and duties is not much different from those of the traditional curator in that the Chief Digital Officer and other professionals in the digital content department select and have an essential role in the construction of mechanisms for interpretation and meaning-making of artworks on a digital display.

Kimberly Drew, former Social Media Manager at The Met (2015–18), said in an interview on 29 April 2017 during the Chicago Humanities Festival that one of her leading roles was to mediate and act as a cultural interpreter and "take super scholarly language and bring it into a world of civilian experience in a way that doesn't water down the scholarship but makes it more approachable". Often, the posts in The Met social media accounts are written in direct speech, using the first person and as informal as any teenager or young adult can be.

Museum social media platforms and, occasionally, museum websites allow visitor interactivity and feedback, enhancing the participation of visitors in the life of the museum and the museum community (Pallas and Economides, 2008). Linda Kelly (2013) argues that social media became relevant in the changing relationship between the museum and audiences and explains how these platforms provide a new space for interaction. In her view, a socio-cultural change, informed by the social, the digital and the participatory, demands a transformative museum that needs to shift to digital communication (Kelly, 2013). The creation of digital content for museum websites and social media platforms introduced a significant change in the way collections are interpreted, presented and communicated to the public.

The digital public of the museum is massively wider, very much beyond the notion of a museum's local or national community, requiring the museum to communicate and present the collections in a more reachable manner, adequate to a global audience. Digital initiatives such as The Met *Connections* contribute to bridging the museum professionals and the digital public. *Connections* is a series of 100 episodes with less than 3 minutes and 30 seconds in which museum staff talk informally about a theme of their choice while showing images from the museum space or the collections. In most cases, museum staff (not only curators) speak in a very informal manner, occasionally sharing personal thoughts and experiences as a form of encouragement to every museum visitor (digital or physical) to appreciate and experience the artworks on display through a personal perspective. This exclusive digital storytelling experience provided through the museum website was produced in 2011 by a team from the digital content department, which included an in-house director, a producer, editors, sound recording technicians, post-production staff and proof-readers, who collaborated with outside production professionals from the industry.

With the establishment of digital content and design departments, museums are increasingly employing a workforce that can bridge culture, social behaviour and digital technologies (Silvaggi and Pesce, 2018). Data collected from employment websites and other websites related to museums and technology (museweb) demonstrates that museums are recruiting digital content managers, digital marketing managers, collections database administrators, directors of community engagement, directors of learning and education strategies, user experience designers, web content managers, social media managers, digital photographers, digital media managers, visual and interaction designers, and many other roles that cross culture, creativity and digital technologies. Job advertisements generally require specific digital technology skills and knowledge, and a bachelor's degree in a related area (art history, museum studies or digital technologies). Museum professionals working in digital content departments frequently hold a degree in art history or museum studies and have proficient digital skills and competences. Staff highly skilled in digital technologies are, and will be, crucial in the digital transformation that museums are going through in the twenty-first century. These emerging professional roles in museums are critical, not only because of their technological significance, but also mostly because of their ability to design human-centred solutions to connect the museum and the public. In a time when the digital becomes more and more relevant in our daily routines, in the ways we engage socially, and in the modes we learn and source information, museums need more professionals acquainted with digital processes of communication.

CURATING NEW MEDIA ART AND NEW MEDIA CURATING ART

Even from a historical perspective, artistic sophistication is commonly related to technological development. Artisans and artists have always sought the integration of technology to materialise and present creative ideas and express conceptual thoughts. The emergence of the new museology approach rapidly embraced the inclusiveness of dynamic and interactive artistic practices involving digital technologies. When dealing with digital art installations, the role of the museum curator is more complex and generally demands active participation of the artist and studio assistants/technicians.

With the transformative forms of contemporary art and the exploration of new media, a new breed of curator has emerged. In the world of contemporary art and new media, the museum curator collaborates directly with the artist(s), who frequently use the gallery space as a white canvas around eccentric concepts of the curator or the artists themselves (Obrist, 2008). As a result, the divide between the role of the artist and the role of the curator becomes gradually indiscernible in which, as argued by some scholars, art curation has become a form of intellectual authorship (Cairns and Birchall, 2013; O'Neil, 2007; 2012). Both the curator and the contemporary artist frequently explore digital technologies as an everyday tool or instrument to communicate and build forms of interaction with the audience. This may result in artistic authorship, curatorial authorship, and sometimes both, when the artist is also a curator or the curator becomes the artist.

These interactions between the digital and artistic/curatorial practice can take place not only in the gallery or museum space, but also, gradually, on the internet, through the use of computer systems in the creative process and the display of artworks on webpages and social media accounts. The use of digital technologies in artistic and curatorial practices constructs a sense of the totality of the exhibition space in the same way it explores conceptual notions of aesthetics of machines, generative art and reality.

As early as the 1950s, inventors/artists such as Desmond Paul Henry, John Whitney and Charles Csuri, among others, started exploring what became known as the computer art movement. Using analogue computer-based drawing machines to form kinetic expressions and motion graphics, interactive computer games and other computer-generated screen media, these inventors/ artists constructed a new bridge between art, innovation and technology. The Whitechapel Art Gallery was one of the instigating institutions exhibiting new media art. In 1956, the gallery showed the exhibition *This is Tomorrow*, an artist/curator collaboration of the Independent Group, a collective of artists, designers, architects and cultural theorists linked to the Institute of Contemporary Arts (ICA) in London. The gallery space was used as a white canvas and was collectively curated by the 38 participants in the exhibition (Blazwick and Yiakoumaki, 2010; Wallis and Alloway, 1987). The exhibition was divided into 12 groups formed by, at least, an architect, a designer, an artist and a theorist, in a combined effort to deploy a new methodology of artistic creation but also exploratory methods of gallery curation. Both the content and narrative of the exhibition discussed the future of integrated new media, technology and mass production in the living space, the built environment, artistic production, and the future of interactivity and user experience because the visitors had no guided interpretation materials.

Over the 1960s and 1970s, several museums and galleries, such as San Francisco MoMA, Walker Art Center, Whitney Museum and ICA, held exhibitions exploring the intersections between art and technology, concerning the ideas of techno-futurism and conceptual art (Gere, 2004). With the development of the information technology and communication industry in the 1980s, most inventors/artists working with technology ended up working in the industry.

It was only at the beginning of the 1990s, with the development of the World Wide Web, that a revival of computer systems art re-emerged.

A mixed exhibition space, between the museum and the internet, evolved in the context of net art and contemporary artistic practice. Initially, museums were eager to integrate digital artistic expressions to capture audiences, who were gradually seduced by the new forms of visualisation using VR, screen media and computer-generated imagery. Some of these examples were the Walker Art Center's *Gallery 9*, San Francisco MoMA's *e-space* and the Whitney Museum's *artport* (Paul, 2006). Similar to early exhibitions involving computer-based systems, the artists are often the curators and technical experts, facilitating an end-to-end artistic and cultural product, generating an increasing democratisation of content and audience participation in selection and filtering.

The technological development over the 1990s and early twenty-first century potentialised the exploration of new artistic and curatorial theoretical approaches, as well as the exploration of technology and computer software as a visual mediator. In this context, the artist and the curator are often versed in technological possibilities enabling the appreciation of digital aesthetics and the development of new ideas rooted in digital cultural theory.

#fuckreality was an exhibition held in October 2018 at Kunstraum Niederoesterreich, an experimental and transdisciplinary exhibition space in Vienna, Austria, in collaboration with the Department of Digital Arts at the University of Applied Arts Vienna. The exhibition was curated by Martin Kusch, Alexandra Schantl and Ruth Schnell, a professor of philosophy and sociology of science, an art historian and curator, and a media artist and curator, respectively. The curators conceptualised an exhibition in which digital technology is viewed from a de-stigmatised perspective, away from media culture (games and films) and commerce. The exhibition rejects the simplistic commercial value and commoditisation of digital technology and points towards an aestheticisation of the notions of the immersive, the real and the virtual in contemporary societies, in which the visitor experiences the dissolution of reality and space.

The use of new media in contemporary artistic practice and curatorial practice demands expertise in various fields, resulting in a high level of collaboration. The paradigm of the collaborative work between master and apprentice in the Renaissance workshop has returned in the contemporary art scene, where the artist works collectively with studio assistants, whose work is providing technical solutions. Artists and the studio assistants then collaborate with the team of curators in the museum or gallery to install, mount and connect artworks. The intangibility of the digital artwork raises several procedural and ethical problems related to lending, reproduction and accessibility that physical artworks do not present. With digital artworks residing in the digital space, museums and galleries are required to secure permissions for links and negotiate the installation of computer systems, which often involves software licences, purchase and rental of hardware, and technical labour (Paul, 2008).

Despite the fact that we are living in a digital age, most art museum institutions are still reluctant to exhibit web and media art, and when they do, the narrative is often static and typically historicises new media art (Graham and Cook, 2010). Science and technology museums, smaller private and independent galleries, or even more often, the internet, are currently the spaces where new media and digital art fit more comfortably. On the one hand, this is a problem of categorisation in which museum institutions regard digital works as technological objects rather than art objects. In the same way, the media specialist is not necessarily regarded as an artist. On the other hand, contemporary art museums play a significant role in the approval, recognition and validation of artists in the art scene. This happens through the prestige of the institution and the accreditation of a team of art experts. However, in a digital age, media artists find mass validation, visibility, acceptance and popularity on the internet, sometimes becoming "influencers" and cyber stars. The success of the internet as a curatorial space became so popular and efficient that it enabled the emergence of other job opportunities and revenue streams for artists using conventional media, such as painting, drawing and illustration, to name a few.

Patreon, a crowdfunding digital platform for creatives, is an excellent example of how the internet, as an alternative virtual repository to display artwork, enables a global scale job market for both digital media and conventional media artists. Creative products and services traded on Patreon and other similar digital platforms are advertised in digital spaces through a process of self-curation. Social media accounts on Facebook, Instagram, Twitter or Tumblr are often used as personal museums or galleries, exhibiting creative work in a way not much different from that of Hans-Ulrich Obrist's "Kitchen Show" in 1991 (Obrist, 2008). This is a form of expanded curation, not necessarily detached from the institutional museum, which Obrist himself is devoted to through his digital gallery in Instagram (@hansulrichobrist). In this digital gallery, he analyses the disappearance of handwriting in the digital age. Obrist collects photos of handwritten notes, sentences, comments and statements by people in his social circle. During his talk at the "me Convention" in Frankfurt on September 2017, Obrist stated that "we can now all do our own museum in a DIY [do it yourself] way".

With the support of digital technologies, the work of curators and artists enters more deeply into social and political spaces because the internet enables participation, collaboration and inclusiveness in artistic and curatorial practices.

NOTE

1.   ORCID iD: 0000-0001-7608-2370, Faculty of Arts and Social Sciences, Universiti Brunei Darussalam (UBD), rui.o.lopes@gmail.com.

REFERENCES

Ambrose, T. and Paine, C. (2006). *Museum basics*. New York, NY: Routledge.

Ball, D. and Earl, C. (2002). Outsourcing and externalisation: Current practice in UK libraries, museums and archives. *Journal of Librarianship and Information Science*, **34**(4), 196–206.

Ball, D., Barton, D., Earl, C. and Dunk, L. (2002). A study of outsourcing and externalisation by libraries with additional reference to the museums and archives domains. Poole: The Council for Museums, Archives and Libraries.

Bautista, S.S. (2014). *Museums in the digital age: Changing meanings of place, community, and culture*. Lanham, MD: Altamira Press.

Bennett, T. (1995). *The birth of the museum. History, theory, politics*. London, UK and New York, NY, USA: Routledge.

Bernstein, S. (2014). GO: Curating with the Brooklyn community. In M. Sanderhoff (ed.), *Sharing is caring: Openness and sharing in the cultural heritage sector* (pp. 186–98). Copenhagen: Statens Museum for Kunst.

Blazwick, I. and Yiakoumaki, N. (eds) (2010). *This is tomorrow*. London: Whitechapel Gallery.

Bowen, J.P. (1997). The World Wide Web and the Virtual Library museums pages. *European Review: Interdisciplinary Journal of the Academia Europaea*, **5**(1), 89–104.

Bowen, J.P. (2000). The virtual museum. *Museum International*, **52**(1), 4–7.

Boylan, P.J. (2008). The museum profession. In S. Macdonald (ed.), *A companion to museum studies* (pp. 415–30). Oxford: Blackwell Publishing.

Cairns, S. and Birchall, D. (2013). Curating the digital world: Past preconceptions, present problems, possible futures. In N. Proctor and R. Cherry (eds), *Museums and the Web 2013*. Silver Spring, MD: Museums and the Web.

Cann, S. (2012). Participatory curatorial practices: An online approach. *International Journal of Humanities and Social Science*, **2**(14), 73–8.

Coffee, K. (2008). Cultural inclusion, exclusion and the formative roles of museums. *Museum Management and Curatorship*, **23**(3), 261–79.

Czegledy, N. (2012). Curatorial models and strategies in a digital age. *Kepes*, **9**(8), 141–55.

Dallas, C. (2004). The presence of visitors in virtual museum exhibitions. Paper presented at Numérisation, Lien Social, Lectures Colloquium, University of Crete, Rethymnon, June.

Falk, J.H. and Dierking, L.D. (2008). Enhancing visitor interaction and learning with mobile technologies. In L. Tallon and K. Walker (eds), *Digital technologies and the museum experience: Handheld guides and other media* (pp. 19–34). Lanham, MD: Altamira Press.

Falk, J.H. and Dierking, L.D. (2018). *Learning from museums*. London: Rowman & Littlefield.

George, A. (2015). *The curator's handbook*. London: Thames and Hudson.

Gere, C. (2004). New media art and the gallery in the digital age. *Tate Papers*, 2. Retrieved from https://www.tate.org.uk/research/publications/tate-papers/02/new -media-art-and-the-gallery-in-the-digital-age.

Giannini, T. and Bowen, J. (2015). A New York museums and Pratt partnership: Building web collections and preparing museum professionals for the digital world. In *MW2015: Museums and the Web 2015*. Retrieved from https://mw2015 .museumsandtheweb.com/paper/a-new-york-museums-and-pratt-partnership -building-web-collections-and-preparing-museum-professionals-for-the-digital -world/.

Giannini, T. and Bowen, J.P. (2019). Rethinking museum exhibitions: Merging physical and digital culture – past to present. In T. Giannini and J.P. Bowen (eds), *Museums and digital culture: New perspectives and research* (pp. 163–93). Cham: Springer.

Graham, B. and Cook, S. (2010). *Rethinking curating: Art after new media*. Cambridge, MA: MIT Press.

Harrison, J. (2000). Outsourcing in museums. *International Journal of Arts Management*, **2**(2), 14–25.

Henning, M. (2008). New media. In S. Macdonald (ed.), *A companion to museum studies* (pp. 302–18). Oxford: Blackwell Publishers.

Jeffery, C. (ed.) (2015). *The artist as curator*. Bristol: Intellect.

Jenkins, H. (2006). *Convergence culture: Where old and new media collide*. New York, NY, USA and London, UK: New York University Press.

Kelly, L. (2007). Visitors and learners: Adult museum visitors' learning identities. In S. Knell, S. MacLeod and S. Watson (eds), *Museum revolutions: How museums changed and are changed*. London, UK and New York, NY, USA: Routledge.

Kelly, L. (2013). The connected museum in the world of social media. In K. Drotner and K. C. Schrøder (eds), *Museum communication and social media: The connected museum* (pp. 54–71). London: Routledge.

Kelly, L. and Russo, A. (2008). From ladders of participation to networks of participation: Social media and museum audiences. In J. Trant and D. Bearman (eds), *Museums and the Web 2008: Proceedings* (pp. 83–92). Toronto: Archives and Museum Informatics.

Laursen, D., Kristiansen, E. and Drotner, K. (2016). The museum foyer as a transformative space of communication. *Nordisk Museologi*, **1**, 69–88.

Lawrenson, A. and O'Reilly, C. (2018). *The rise of the must-see exhibition: Blockbusters in Australian museums and galleries*. London, UK and New York, NY, USA: Routledge.

O'Neil, P. (2007). The curatorial turn. In J. Rugg and M. Sedgwick (eds), *Issues in curating contemporary art and performance* (pp. 13–28). Bristol: Intellect.

O'Neil, P. (2012). *The culture of curating and the curating of culture(s)*. Cambridge, MA: MIT Press.

Obrist, H.U. (2008). *A brief history of curating*. Zurich: JRP/Ringier.

Pallas, J. and Economides, A.A. (2008). Evaluation of art museums' web sites worldwide. *Information Services and Use*, **28**(1), 45–57.

Patel, M., White, M., Walczak, K. and Sayd, P. (2003). Digitisation to presentation: Building virtual museum exhibitions. Paper presented at Vision, Video and Graphics 2003, Bath, UK.

Paul, C. (2006). Flexible contexts, democratic filtering, and computer-aided curating: Models for online curatorial practice. In J. Krysa (ed.), *Curating, immateriality, systems: On curating digital media* (pp. 85–105). New York, NY: Autonomedia Press.

Paul, C. (2008). Challenges for a ubiquitous museum: From the white cube to the black box and beyond. In C. Paul (ed.), *New media in the white cube and beyond: Curatorial models for digital art* (pp. 53–75). Berkeley, CA: University of California Press.

Sabharwal, A. (2015). *Digital curation in the digital humanities: Preserving and promoting archival and special collections.* Waltham, MA: Chandos Publishing.

Samis, P. (2008). The exploded museum. In L. Tallon and K. Walker (eds), *Digital technologies and the museum experience: Handheld guides and other media* (pp. 3–18). Lanham, MD: Altamira Press.

Schorch, P. (2013). The experience of a museum space. *Museum Management and Curatorship*, **28**(2), 193–208.

Silvaggi, A. and Pesce, F. (2018). Job profiles for museums in the digital era: Research conducted in Portugal, Italy, and Greece within the Mu.SA project. *ENCATC Journal of Cultural Management & Policy*, **8**(1), 56–69.

Tallon, L. (2017). Digital is more than a department, it is a collective responsibility (Blog post). Retrieved from https://www.metmuseum.org/blogs/now-at-the-met/2017/digital-future-at-the-met.

Vergo, P. (ed.). (1989). *The new museology.* London: Reaktion Books.

Walczak, K., Cellary, W. and White, M. (2006). Virtual museum exhibitions. *IEEE Computer*, **39**(3), 93–5.

Walker, K. (2008). Structuring visitor participation. In L. Tallon and K. Walker (eds), *Digital technologies and the museum experience: Handheld guides and other media* (pp. 109–24). Lanham, MD: Altamira Press.

Wallis, B. and Alloway, L. (1987). *This is tomorrow today: The Independent Group and British pop art.* New York, NY: P.S.1, Institute for Art and Urban Resources.

Wojciechowski, R., Walczak, K., White, M. and Cellary, W. (2004). Building virtual and augmented reality museum exhibitions. In *Proceedings of the ninth international conference on 3D Web technology* (pp. 135–44). New York, NY: ACM.

PART III

Changing contexts of creative work

9. The role of casual creative environments for creative work in cities: implications for the future creative city[1]

Ana Bilandzic,[2] Onur Mengi[3] and Greg Hearn[4]

INTRODUCTION

The rise of the intangible economy (Haskell and Westlake, 2018) has reshaped the deployment of human capital and the ways it functions as an essential input into innovation. As Hearn and McCutcheon (2020) point out, this partly explains why creative services work is growing rapidly. As such, human capital has also been listed as the most common factor in innovation indexes for cities (Mulas et al., 2016).

According to Florida et al. (2017): "Place has come to replace the industrial corporation as the key economic and social organizing unit of our time. Cities are not just containers for smart people: they are the enabling infrastructure where connections take place, networks are built and innovative combinations are consummated" (p. 92).

Key to this enabling infrastructure are what we term *casual creative environments* (CCEs) (Bilandzic et al., 2020). The popularity of CCEs as new spaces for idea generation, informal creativity and knowledge spillovers is suggested by their growing numbers all over the world (Bouncken and Reuschl, 2016; Kojo and Nenonen, 2014). CCEs may take the form of coworking spaces (Garrett et al., 2017), innovation spaces (Bilandzic et al., 2018) or creative workplace transformations such as at Google and Facebook (Mahlberg and Riemer, 2017). The adjective *casual* encapsulates the idea of freedom: freedom of association and freedom of choice about whether to be part of a particular space or not. And freedom is a key component of Deci's (1972) idea of intrinsic motivation, which is important to discretionary effort and activity. CCEs should therefore be thought of as bottom-up rather than top-down organising routines. They provide workspaces for freelancers, mobile creative workers, small companies and fledgling start-ups. Therefore, an analysis of CCEs within the innovation ecosystem of a city could provide valuable insights for understanding creative work and its role in innovation in city contexts in the future.

We use Brisbane, Australia, as a case to examine the emergence of CCEs. McKinsey and Company and the World Economic Forum (cited in The New Climate Economy, 2014) categorised innovation clusters based on their growth momentum of United States patents granted between 1997 and 2006, and captured their diversity based on the numbers of separate companies and patent sectors within the cluster in 2006. They classified Brisbane as a *hot spring* that is a "small, fast-growing hub" (The New Climate Economy, 2014, p. 16). Brisbane's former Lord Mayor Graham Quirk shared numbers on the city's wide diversity of people, including 78 000 international students from 160 countries (Choose Brisbane, 2017, para. 3–4), who add to the 1.2 million population in Brisbane's local government area (LGA) (Australian Bureau of Statistics, 2018). Digital nomads are another key part of this ideation ecosystem because their global networks enable the development of locally produced, globally relevant capabilities. We suggest this cultural diversity and know-how can add to a city's innovation resources if planned and managed with careful consideration.

The key and relatively new emergence of CCEs in cities' innovation landscapes motivates our analysis of CCEs' role within the innovation ecosystem of cities. We have two aims in undertaking this analysis: to understand the rapid emergence of CCEs over time; and to learn about their geographical distribution, spatial characteristics, and purpose within the city. Then, we examine the implications of these findings for the future of creative work, and creative cities in future. Our findings provide policy makers, urban planners, and CCE owners and managers with guidelines for informed decision-making that consider creative workers' needs for future creative work spaces and encourage the interplay between different stakeholders in the innovation ecosystems.

CCES AS NEW CREATIVITY AND INNOVATION SPACES

Hearn and McCutcheon (2020) argue that the twin drivers of technological change and rise in intangible capital are changing the institutions and systems that govern and enable creative work in profound ways. Jobs heavy in foot-loose creativity and knowledge are out-pacing jobs anchored in equipment and factories, and this is changing modes and places of work. In parallel, much critical consideration has been given to the impact of such trends on cities (e.g., Mayer, 2013). Urban planning must consider new spaces for cultural production and knowledge-based activities (Mayer, 2013; Pancholi et al.,

2017). Oksanen and Hautamäki (2014) describe innovation ecosystems as local hubs of global networks or technology platforms that consist of particular actors and dynamic processes, which together produce solutions to different challenges. As a concept, innovation ecosystems also have their roots in industry and business clusters (Estrin, 2008; Porter, 1998) where knowledge-based activities occur (Audretsch, 2002). Winden et al. (2012) suggest that as an ecosystem grows with success, it will draw in new firms and talented people from the outside. However, they suggest ecosystems can only function in places that provide quality of life, accessibility and cultural assets to attract creative people.

A broad range of creative spaces in such ecosystems have emerged to meet the demands of creatives and knowledge workers who are seen as crucial actors in the service-centred knowledge economy. These spaces are defined as any of the following:

- Innovation spaces (Bilandzic et al., 2018; Casadevall et al., 2018), coworking spaces (Bilandzic et al., 2013; Capdevila, 2015; Garrett et al., 2017; Kojo and Nenonen, 2014; Mahlberg and Riemer, 2017) and CCEs (Bilandzic et al., 2020).
- Knowledge and innovation spaces (Evers et al., 2010; Lönnqvist et al., 2014; Pancholi et al., 2017; Yigitcanlar and Bulu, 2016).
- Innovation clusters (Esmaeilpoorarabi et al., 2018; Kiuru and Inkinen, 2017), for example, science precincts, technology clusters or techno-industrial districts (Huggins, 2008).

Creativity and knowledge spaces are a component of innovation ecosystems that are emerging in major cities around the world (Mulas et al., 2016). These spaces comprise various types of spatial and organisational structures.

SPATIAL TYPES OF CCES

Knowledge workers, including creatives, report that they use CCEs as alternative working environments that expose them to opportunities for networking, collaboration, coworking and being part of a community (Bouncken and Reuschl, 2016; Czarniawska, 2014; Foertsch, 2011; Garrett et al., 2017), or as places for informal learning that allow perpetual messiness for new skill acquisition and tinkering (Bilandzic and Foth, 2013). Being bottom-up and casual, these spaces help and mobilise their users, and could be seen as precursors to innovation (Bilandzic et al., 2018) because users get support even before starting up their business (PTI, 2015). Nevertheless, users of CCEs report that they struggle finding a suitable CCE for their work (Forlano, 2011) and dislike being locked in an echo chamber inhabited only by like-minded people

(Bilandzic et al., 2018). Bilandzic et al. (2018) and Foth et al. (2017) identified CCE work and identity cultures that excluded some workers, particularly those outside the mainstream of commercial start-up culture, and argued for a more inclusive approach to innovation from policy makers and CCE managers. Research has discussed the need for different spatial configurations along the creativity process (Kristensen, 2004), as well as the tension coming from different user groups working in a standardised global model of start-up culture and workspace (Foth et al., 2016).

To concretise our investigation of such issues, we proposed three spatial types of CCEs, namely: fixed, flexible and free.

Fixed CCEs are spaces equipped with necessary facilities, equipment and machinery. They provide their users with hot-desks, hirable offices, meeting rooms or workshops/ateliers with 3D printers, laser-cutters and working desks that are physically fixed in the facility. These CCEs are generally seen as spaces to rent in corporate business environments that may be user-appropriated, or workshops for rent more for specialised work purposes that provide rigid and immobile working environments (Foth et al., 2016).

Flexible CCEs have a configurable spatial layout compared to the fixed CCEs and may also include equipment. The use of the space depends on the users' preferences, and how they can use the existing spatial configuration and adapt it for particular creative work.

Free CCEs are unoccupied and empty spaces that are spatially free for users or customers to navigate, furnish and design inside. These CCEs are often found in the form of pavilions and post-industrial spaces used for pop-up workshops or short-term events.

In addition to how spatial configuration of CCEs can constrain motivation and diversity, we also explored the need for different spatial configurations along the innovation process (Kristensen, 2004). We therefore also classified the purpose of CCEs according to whether they were for: *education and learning*; *tinkering*; *art and design production*; *meetup and networking*; *independent coworking*; *art and design exhibition*; and *retail*. These purposes are all implicated in creative work as it manifests itself in the creative economy, as described by Hearn and McCutcheon (2020) and Hearn et al. (2020). This allowed us to investigate spatial configurations of CCEs in relation to their purposes (Kristensen, 2004; Schmidt et al., 2014), as well as their business entity type.

METHODOLOGY

We identified CCEs in Brisbane through online desktop research accessing publicly available data, following Casadevall et al.'s (2018) finding that this is how users search and find informal places if they were not recommended to them through offline word-of-mouth or Facebook advertisements. We identified six online platforms that inform or allow users to browse and hire CCEs by searching online for related keywords that are a combination of *innovation spaces*, *creative spaces*, *coworking spaces* and *Brisbane*, or similar terms:

- coworker.com;
- spacely.com.au;
- workfrom.co;
- theurbanlist.com/brisbane/a-list/best-creative-workshops-brisbane;
- creativespaces.net.au;
- weekendnotes.com.

We identified 133 spaces that fit our definition of a CCE and that were operating during data collection in November 2018. Therefore, we excluded CCEs that might have operated in the past, but closed their business before November 2018. We collated CCEs' names and websites and used this information to find out their locational coordinates identified using Google Maps. This data was uploaded to the Google "MyMap" feature, creating a geographical map of the CCEs. Eighty per cent of the sample were clustered around the central business district (CBD) and immediately surrounding suburbs. The rest were distributed in all directions in mid-range to outer suburbs.

Further, we added information for all CCEs' business commencement year and business entity type, via the Australian Business Register (ABR) website (abr.business.gov.au) (Australian Government, n.d.). The commencement year of CCEs is an indicator of perceived demand for these types of spaces. The business entity type is coded by the ABR according to their tax obligations and includes every legal structure possible in Australia (e.g., private/public companies, sole traders, trusts, cooperatives and charities). Based on the CCEs' visual and descriptive information available on their websites, we then documented their spatial configuration type (i.e., free, flexible or fixed), as well as their primary and secondary purpose (i.e., education and learning; tinkering; art and design production; meetup and networking; independent coworking; art and design exhibition; and retail).

In the next section, we present and discuss our findings in the following order: number of CCEs emerging over time; CCEs' geographical distribution based on their spatial configuration and purpose; and the relation of CCEs' spatial configuration to their purposes and business entity types.

CCES' SPATIAL AND TEMPORAL DISTRIBUTION

The 133 CCEs and collected information about them are summarised in Table 9.1. The numbers in brackets indicate the number of CCEs classified under the

Table 9.1   *Summary of data collected on 133 CCEs in Brisbane*

CCE types	Primary purposes	Secondary purposes	Business entity types	Business commencement year
Fixed (69)	Independent	Meetup &	Private Company (59)	1891 (1)
Flexible (47)	coworking (53)	networking (40)	Individual/Sole Trader	1970 (1)
Free (17)	Art & design	Education &	(17)	1999 (8)
	production (35)	learning (27)	Trusts (14)	2000 (22)
	Education &	Art & design	Other Incorporated	2001 (1)
	learning (15)	production (14)	Entity (11)	2003 (2)
	Meetup &	Art & design	Other Unincorporated	2004 (1)
	networking (15)	exhibition (7)	Entity (4)	2005 (3)
	Retail (9)	Tinkering (5)	Public Company (4)	2006 (3)
	Art & design	Independent	State Government	2007 (2)
	exhibition (6)	coworking (3)	Entity (3)	2008 (17)
	Tinkering (0)	Retail (0)	Others (3)	2009 (2)
		N/A (37)	N/A (18)	2010 (6)
				2011 (6)
				2012 (5)
				2013 (7)
				2014 (6)
				2015 (13)
				2016 (7)
				2017 (12)
				N/A (8)

Source: Includes data from the Australian Business Register (Australian Government, n.d.).

respective classification. We determined 69 *fixed CCEs*, 47 *flexible CCEs* and 17 *free CCEs*. The most frequent primary purpose was *independent coworking* assigned to 53 CCEs, followed by 35 CCEs primarily focusing on *art and design production*, and 15 CCEs both on *education and learning*, and *meetup and networking*. Among the secondary purpose, we found 40 CCEs supporting *meetup and networking*, followed by 37 *N/A* classifications because the second purpose for those was uncertain. The most common business entity type was private companies, followed by individual/sole traders and trusts. Most CCEs commenced after the year 2000.

The majority of CCEs are Australian private companies located in fixed or flexible facilities and providing space most often for *independent coworking* or *art and design production*. Three CCEs are state-government owned. Each is located in a different spatial configuration type. Private companies and individual or sole traders predominantly own CCEs supporting *art and design production*. Similarly, individuals and sole traders are the most likely to own

CCEs with an *education and learning* focus. We also observe one CCE being a cooperative: Reverse Garbage, which collects waste and offers disposal services and workshops and talks on topics about the environment and waste. In Queensland, cooperatives get involved in economic and social activities (Co-operative Development Services, 2019). This characteristic makes this CCE unique, being the sole CCE in our data set entailing social care and responsibility defined by its business entity type.

Emergence of CCEs over Time

Kojo and Nenonen (2014) investigated such spaces from the 1960s, showing that coworking spaces went through an evolution rather than being established overnight. As per Table 9.1, all but ten CCEs in Brisbane commenced their business in the new millennium. We found spikes of CCE openings for the years 2000, 2008, 2015 and 2017, compared with an almost steady rate of 1 to 8 openings with few variations over the timeline for all other years. We also found that while the number of commencements of both fixed and flexible CCEs has grown across the time period, the growth in relative popularity of flexible CCEs is notable. The openings of free CCEs were stable over time, with 2 to 4 business registrations per time period.

The overall popularity of CCEs in Brisbane since 2000 could be explained by:

- The global trend of such spaces (Bouncken and Reuschl, 2016), which is enabled through the internet to telework (Kojo and Nenonen, 2014) and workers becoming digital nomads (Czarniawska, 2014; Liegl, 2014; Rossitto et al., 2014).
- The trend to entrepreneurship and freelance work. In Foertsch's (2011) survey of 661 coworkers from 24 countries, most coworkers (54 per cent) were freelancers, 20 per cent were entrepreneurs who employed others, and 20 per cent were permanently employed.
- Brisbane City Council's (BCC's) funding and policy supporting openings of coworking spaces since 2007 and accelerated support from 2013. For example, the *Brisbane Economic Development Plan 2012–2031* (BCC, 2012), *Creative Brisbane Creative Economy 2013–22* (BCC, 2013), and the *Brisbane 2022 New World City Action Plan* (Lord Mayor's Economic Development Steering Committee, 2015).
- Brisbane's multicultural population, including 78 000 international students from 160 countries (Choose Brisbane, 2017, para. 3–4) who form part of the entrepreneurial crowd.

• The co-existence and accessibility in Brisbane of critical infrastructure and stakeholders in innovation (e.g., universities, private research centres, angel investors, learning facilities for digital skills).

Geographical Distribution of CCEs

Noting that most CCEs were located in and around the CBD, we concentrated on this region for the forthcoming figures. Figure 9.1 shows the distribution of CCEs based on their spatial configuration type. With this more fine-grained focus, we see a belt of CCEs that stretches from residential suburbs just north-east and south-west of the CBD, both of which have been evolving as extensions of retail, entertainment and commercial business districts in the CBD. Free CCEs tend to be scattered in suburbs near the CBD, rather than in the CBD. In contrast, fixed CCEs cluster along the retail/entertainment/commercial belt from the north-east through the CBD to the south-west belt, but are highly concentrated in the CBD. Bundles of flexible CCEs are also found in this belt, but are notably also scattered in suburbs all around the CBD.

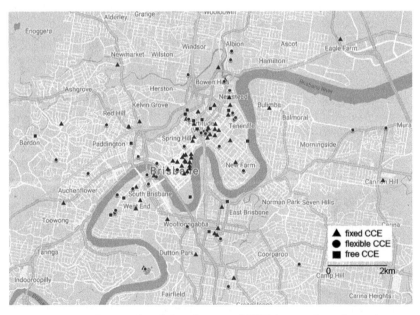

Figure 9.1   Geographical distribution of CCEs' spatial configuration types in Brisbane's CBD and neighbouring suburbs

We also investigated the CCEs' location based on their primary purpose. A total of 53 *independent coworking* and 35 *art and design production* CCEs form the two biggest purpose groups and compose 66 per cent of all CCEs. *Independent coworking* opportunities cluster around the CBD along the east–west orientation; however, *art and design production* spaces are found more frequently in the surrounding suburbs as a kind of geographical *creative fringe* (Bilandzic et al., 2018; Casadevall et al., 2018) situated almost completely outside of the CBD. We also found that none of the CCEs focused on providing primarily *tinkering* opportunities, pointing to a lack of skunkworks (Foth, 2015) and the perpetual and experimental messiness (Bilandzic and Foth, 2013) needed for creative experimentation.

Common Purpose Relations Between CCEs

In order to further understand the functioning of the CCEs holistically in the innovation ecosystem of Brisbane, we now turn attention to relations between CCEs in terms of purpose. This section is based on examining *common purpose* links between CCEs using data on their primary and secondary purposes, spatial configuration types and business entity types. Figure 9.2 is depicted as a network of common purposes differentiated with respect to fixed, flexible and free space. The biggest hub in the network is formed by fixed or flexible CCEs for the primary purpose of *independent coworking*. Further, the majority of these CCEs have *meetup and networking* as their secondary purpose. CCEs with a *meetup and networking* primary purpose are more likely to be in flexible or free spaces where users have more control over the configuration of the space, which may be useful for certain events.

CCEs that have *art and design production* as their primary purpose include all three spatial configuration types, though those in flexible spaces form the second-biggest cluster in the network. Their secondary purpose is most often *education and learning*. Similarly, spaces that focus primarily on *education and learning* include all three CCE types and have *art and design production* as their secondary purpose. The tight relation between *art and design* and *education and learning* suggests capabilities, skills and talents that can be recruited to particular ideas. Figure 9.2 also shows the types of CCEs that connect the two main clusters or the two most frequent primary purposes. Referring back to the raw data, we found these in fact were all incubators with a secondary function in *art and design*. Art and design is a key element of the ability to articulate creative ideas in a visual form that is easy to understand. It may be that these CCEs provide key capabilities in translating abstract ideas into intelligible, communicable designs or visual representations.

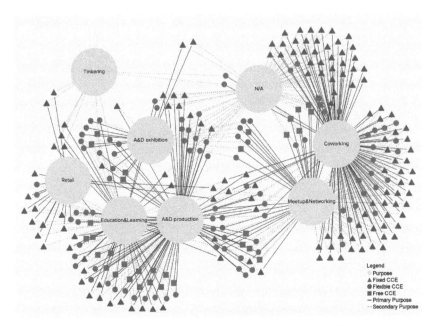

Figure 9.2   CCEs' relational configuration according to their two
                 primary purposes

THE FUTURE OF CREATIVE WORK AND CCES

Much has been written about *creative cities* and their relevance to the creative
economy (e.g., Florida et al., 2017; Fuente and van Luyn, 2020; Scott, 2006;
Storper and Scott, 2009). In creative cities, CCEs have been proliferating,
and are an integral part of the innovation ecosystem for creatives and other
knowledge workers who pursue their new entrepreneurial and innovative
endeavours through co-creation, coworking and collaboration (DeGuzman and
Tang, 2011; Hughes et al., 2014). Our findings are consistent with the global
trend of growing numbers of CCEs (Bouncken and Reuschl, 2016) and suggest
that these spaces are important contexts for the future of creative work (Florida
et al., 2017). Emerging creativity and knowledge-intensive spaces evolve in
various ways (e.g., top-down and bottom-up initiatives), but their social heter-
ogeneity, and the links and interactions between individuals, places and other
companies are considered to be major economic drivers (Scott, 2006; Storper
and Scott, 2009). However, their economic significance should not be thought
of narrowly, solely in terms of commercial start-ups and immediate regional
impact. Granovetter (2005) reminds us that economic activity is embedded in

social structures, which establish important preconditions for economic activity, such as trustworthy norms, novel ideational flows, and shared cost efficiencies through access to non-economic institutions and networks. Ideation, social support for ideation, and social innovations per se are key functions of *underground epistemic communities* that underlie creative cities (Cohendet et al., 2010).

Our study of Brisbane's evolving innovation ecosystem found a diversity of purpose for CCEs that reflect both social precursors to innovation as well as current commercially driven agendas. This diversity is captured in the primary purpose of the CCEs: education & learning (15); A&D production (35); meetup & networking (15); independent coworking (53); A&D exhibition (6); and retail (9). This spectrum provides a mix of capacity built from skill development to social facilitation, business support, commercial activity and engagement with consumers and audience. In terms of creative workers, rather than knowledge workers in general, the Brisbane CCEs are utilised by a wide range of creative occupations, including creative services work such as design and software and digital content, as well as artists in the creative fringe (see Hearn and McCutcheon, 2020).

Given the growing popularity of CCEs for creative work, we argue that in the future CCEs need more thoughtful incorporation within the dynamics of cities' innovation ecosystems, with regard to locational choices, purposes for user engagement and entity legal structures (public, business or non-governmental organisation). CCEs would benefit city innovation ecosystems that facilitate a high concentration of people and companies in close proximity that together advance innovation holistically (Athey et al., 2007). These considerations could engage *diverse* human capital to create novel products and services in creative and other industries (Hearn and McCutcheon, 2020) and optimise participation in global value chains (Hearn and Pace, 2006). This will have a net positive impact on creative employment options (Hearn and McCutcheon, 2020). In line with this, CCE owners and managers may consider strategies to attract people from different socio-economic, cultural and educational backgrounds to create diversity in their space that fosters innovation (Amelie, 2017; Bilandzic et al., 2018; Johnstone et al., 2016). To nurture and build a comprehensive and diverse innovation ecosystem, urban planners should consider strategies that are inclusive of all creatives and knowledge workers, and their need for spaces, specifically CCEs, in relevant locations that are conducive to the nature of their work. In addition to that, urban policies that take into account the organic development of these spaces might also help to achieve innovative environments for creative work. Future research may investigate the relationships between CCEs and other stakeholders (e.g., universities, cultural places, and private research and development institutions) in the city's innovation ecosystem to get a more holistic understanding of the ecosystem and potential collaborations within it. By avoiding a narrow economistic perspective that leads to monocultures of values and skills, CCEs can expand the opportunities for creative work in the future, whether that work is dedicated to artistic excellence, social impact, or a prosperous and just economy.

NOTES

1. Adapted from Bilandzic, A., Mengi, O. and Hearn, G. (2019). Spatial and temporal emergence of casual creative environments in Brisbane. In I. Bitran, S. Conn, C. Gernreich, M. Heber, K. Huizingh, O. Kokshagina, M. Torkkeli et al. *Proceedings of The XXX ISPIM Innovation Conference – Celebrating Innovation: 500 Years Since da Vinci*, June. Florence, Italy.
2. ORCID iD: 0000-0002-0746-0590, Creative Industries Faculty, Queensland University of Technology, ana.bilandzic@qut.edu.au.
3. ORCID iD: 0000-0002-0598-9298.
4. ORCID iD: 0000-0003-2245-3433.

REFERENCES

Amelie, M. (2017). What Google and King does to achieve more diversity. *The Wayback Machine*, 26 May. Retrieved from https://web.archive.org/web/20170913085743/ https://planet.telenor.com/2017/05/26/what-google-and-king-does-to-achieve-more -diversity/.

Athey, G., Glossop, C., Harrison, B., Nathan, M. and Webber, C. (2007). *Innovation and the city: How innovation has developed in five city-regions*. London: National Endowment for Science, Technology and the Arts (Nesta).

Audretsch, D.B. (2002). The innovative advantage of US cities. *European Planning Studies*, **10**(2), 165–76.

Australian Bureau of Statistics (2018). Data by region. Retrieved from http://stat.abs .gov.au/itt/r.jsp?databyregion#/. (Last accessed 19 September 2018.)

Australian Government (n.d.). ABN lookup. Retrieved from https://abr.business.gov .au/Help/Glossary#tradingname. (Last accessed 8 May 2019.)

Bilandzic, A., Casadevall, D., Foth, M. and Hearn, G. (2018). Social and spatial precursors to innovation: The diversity advantage of the creative fringe. *The Journal of Community Informatics*, **14**(1), 160–82.

Bilandzic, M. and Foth, M. (2013). Libraries as coworking spaces: Understanding user motivations and perceived barriers to social learning. *Library Hi Tech*, **31**(2), 254–73.

Bilandzic, A., Foth, M. and Hearn, G. (2020). The role of fab labs and living labs for economic development of regional Australia. In E. Fuente and A. Van Luyn (eds), *Regional cultures, economies and creativity: Innovating through place in Australia and beyond*. Abingdon: Routledge.

Bilandzic, A., Mengi, O. and Hearn, G. (2019). Spatial and temporal emergence of casual creative environments in Brisbane. In I. Bitran, S. Conn, C. Gernreich, M. Heber, K. Huizingh, O. Kokshagina, M. Torkkeli, et al. (eds), *Proceedings of The XXX ISPIM Innovation Conference – Celebrating Innovation: 500 Years Since da Vinci*, June. Florence, Italy.

Bilandzic, M., Schroeter, R. and Foth, M. (2013). Gelatine: Making coworking places gel for better collaboration and social learning. In H. Shen, R. Smith, J. Paay, P. Calder and T. Wyeld (eds), *Proceedings of the 25th Australian Computer–Human Interaction Conference: Augmentation, Application, Innovation, Collaboration* (pp. 427–36). New York, NY: ACM.

Bouncken, R.B. and Reuschl, A.J. (2016). Coworking-spaces: How a phenomenon of the sharing economy builds a novel trend for the workplace and for entrepreneurship. *Review of Managerial Science*, **12**(1), 317–34.

Brisbane City Council (BCC) (2012). Brisbane's economic development plan 2012–2031. Retrieved from https://www.brisbane.qld.gov.au/about-council/governance-strategy/business-brisbane/growing-brisbanes-economy/brisbanes -economic-development-plan.

Brisbane City Council (BCC) (2013). Creative Brisbane creative economy 2013–22. Retrieved from https://www.brisbane.qld.gov.au/sites/default/files/creative _brisbane_creative_economy_2013-2022.pdf. (Last accessed 21 November 2018.)

Capdevila, I. (2015). Co-working spaces and the localised dynamics of innovation in Barcelona. *International Journal of Innovation and Technology Management*, **19**(3). Retrieved from https://papers.ssrn.com/sol3/papers.cfm?abstract_id=2502813.

Casadevall, D., Foth, M. and Bilandzic, A. (2018). Skunkworks finder: Unlocking the diversity advantage of urban innovation ecosystems. In *Proceedings of the 30th Australian Conference on Computer–Human Interaction* (pp. 145–55). New York, NY: ACM.

Choose Brisbane (2017). Brisbane international student numbers on the up. Retrieved from https://www.choosebrisbane.com.au/study/news-and-events/news/brisbane -international-student-increase-numbers. (Last accessed 26 October 2019.)

Cohendet, P., Grandadam, D. and Simon, L. (2010). The anatomy of the creative city. *Industry and Innovation*, **17**(1), 91–111.

Co-operative Development Services (2019). Queensland co-operatives. Retrieved from https://www.coopdevelopment.org.au/qldlinks.html. (Last accessed 9 May 2019.)

Czarniawska, B. (2014). Nomadic work as life-story plot. *Computer Supported Cooperative Work*, **23**(2), 205–21.

Deci, E.L. (1972). Intrinsic motivation, extrinsic reinforcement, and inequity. *Journal of Personality and Social Psychology*, **22**(1), 113–20.

DeGuzman, G.V. and Tang, A.I. (2011). *Working in the unoffice: A guide to coworking for indie workers, small businesses, and nonprofits.* San Francisco, CA: Night Owls Press LLC.

Esmaeilpoorarabi, N., Yigitcanlar, T. and Guaralda, M. (2018). Place quality in innovation clusters: An empirical analysis of global best practices from Singapore, Helsinki, New York, and Sydney. *Cities*, **74**, 156–68.

Estrin, J. (2008). *Closing the innovation gap: Reigniting the spark of creativity in a global economy.* New York, NY: McGraw-Hill Education.

Evers, H., Gerke, S. and Menkhoff, T. (2010). Knowledge clusters and knowledge hubs: Designing epistemic landscapes for development. *Journal of Knowledge Management*, **14**(5), 678–89.

Florida, R., Adler, P. and Mellander, C. (2017). The city as innovation machine. *Regional Studies*, **51**(1), 86–96.

Foertsch, C. (2011). The coworker's profile. *Deskmag*, 13 January. Retrieved from http://www.deskmag.com/en/the-coworkers-global-coworking-survey-168. (Last accessed 9 March 2018.)

Forlano, L. (2011). Building the open source city: Changing work environments for collaboration and innovation. In M. Foth, J. Donath, L. Forlano, M. Gibbs and C. Satchell (eds), *From social butterfly to engaged citizen* (pp. 437–60). Cambridge, MA: MIT Press.

Foth, M. (2015). Australia needs an innovation "skunkworks". *The Conversation*, 2 December. Retrieved from https://theconversation.com/australia-needs-an -innovation-skunkworks-51326. (Last accessed 5 December 2017.)

Foth, M., Forlano L. and Bilandzic M. (2016). Mapping new work practices in the smart city. In H. Friese, G. Rebane, M. Nolden and M. Schreiter (eds), *Handbuch Soziale Praktiken und Digitale Alltagswelten* (pp. 1–13). Wiesbaden: Springer.

Foth, M., Hughes, H.E., Dezuanni, M.L. and Mallan, K. (2017). *Skunkworks: Designing regional innovation spaces for the creative fringe.* Paper presented at Designing Participation for the Digital Fringe, Troyes, France.

Fuente, E. and Van Luyn, A. (eds) (2020). *Regional cultures, economies and creativity: Innovating through place in Australia and beyond.* Abingdon: Routledge.

Garrett, L.E., Spreitzer, G.M. and Bacevice, P.A. (2017). Co-constructing a sense of community at work: The emergence of community in coworking spaces. *Organization Studies*, **38**(6), 821–42.

Granovetter, M. (2005). The impact of social structure on economic outcomes. *Journal of Economic Perspectives*, **19**(1) 33–50.

Haskell, J. and Westlake, S. (2018). *Capitalism without capital: The rise of the intangible economy.* Princeton, NJ: Princeton University Press.

Hearn, G. and Pace, C. (2006). Value-creating ecologies: Understanding next generation business systems. *Foresight*, **8**(1), 55–65.

Hearn, G., Cunningham, S., McCutcheon, M. and Ryan, M. (2020). The relationship between creative employment and local economies outside capital cities. In G. Hearn (ed.), *The future of creative work: Creativity and digital disruption.* Cheltenham, UK and Northampton, MA, USA: Edward Elgar Publishing, pp. 34–56.

Hearn, G. and McCutcheon, M. (2020). The creative economy: the rise and risks of intangible capital and the future of creative work. In G. Hearn (ed.), *The future of creative work: Creativity and digital disruption.* Cheltenham, UK and Northampton, MA, USA: Edward Elgar Publishing, pp. 14–33.

Huggins, R. (2008). The evolution of knowledge clusters: Progress and policy. *Economic Development Quarterly*, **22**(4), 277–89.

Hughes, M., Morgan, R.E., Ireland, R.D. and Hughes, P. (2014). Social capital and learning advantages: A problem of absorptive capacity. *Strategic Entrepreneurship Journal*, **8**(3), 214–33.

Johnstone, S., Choi, J.H-J. and Leong, J. (2016). Designing for diversity: Connecting people, places, and technologies in creative community hubs. In *Proceedings of the 28th Australian Conference on Computer–Human Interaction* (pp. 135–9). New York, NY: ACM.

Kiuru, J. and Inkinen, T. (2017). Predicting innovative growth and demand with proximate human capital: A case study of the Helsinki metropolitan area. *Cities*, **64**, 9–17.

Kojo, I. and Nenonen, S. (2014). Evolution of co-working places: Drivers and possibilities. *Intelligent Buildings International*, 1–12.

Kristensen, T. (2004). The physical context of creativity. *Creativity and Innovation Management*, **13**(2), 89–96.

Liegl, M. (2014). Nomadicity and the care of place: On the aesthetic and affective organization of space in freelance creative work. *Computer Supported Cooperative Work*, **23**(2), 163–83.

Lönnqvist, A., Käpylä, J., Salonius, H. and Yigitcanlar, T. (2014). Knowledge that matters: Identifying regional knowledge assets of the Tampere region. *European Planning Studies*, **22**(10), 2011–29.

Lord Mayor's Economic Development Steering Committee (2015). Brisbane 2022 new world city action plan. Brisbane, QLD: Brisbane City Council.

Mahlberg, T. and Riemer, K. (2017). Coworking spaces Australia: The new places where people work, businesses grow, and corporates connect. Retrieved from Sydney Business Insights website: http://sbi.sydney.edu.au/coworking-spaces-aust.

Mayer, M. (2013). First world urban activism: Beyond austerity urbanism and creative city politics. *Cityscape*, **17**(1), 5–19.

Mulas, V., Minges, M. and Applebaum, H. (2016). Boosting tech innovation: Ecosystems in cities: A framework for growth and sustainability of urban tech innovation ecosystems. *Innovations: Technology, Governance, Globalization*, **11**(1–2), 98–125.

Oksanen, K. and Hautamäki, A. (2014). Transforming regions into innovation ecosystems: A model for renewing local industrial structures. *Innovation Journal*, **19**(2), Article 5.

Pancholi, S., Yigitcanlar, T. and Guaralda, M. (2017). Place making for innovation and knowledge-intensive activities: The Australian experience. *Technological Forecasting and Social Change*, **146**, 616–25.

Porter, M.E. (1998). Clusters and the new economics of competition. *Harvard Business Review*, **76**(6), 77–90.

PTI (2015). PM Narendra Modi announces "Start up; Stand up India" initiative to create more jobs. *The Indian Express*, 15 August. Retrieved from http://indianexpress.com/article/india/india-others/pm-narendra-modi-announces-start-up-stand-up-india -initiative-to-create-more-jobs/. (Last accessed 31 January 2018.)

Rossitto, C., Bogdan, C. and Severinson-Eklundh, K. (2014). Understanding constellations of technologies in use in a collaborative nomadic setting. *Computer Supported Cooperative Work*, **23**(2), 137–61.

Schmidt, S., Brinks, V. and Brinkhoff, S. (2014). Innovation and creativity labs in Berlin. *The German Journal of Economic Geography*, **58**(1), 232–47.

Scott, A.J. (2006). Creative cities: Conceptual issues and policy questions. *Journal of Urban Affairs*, **28**(1), 1–17.

Storper, M. and Scott, A.J. (2009). Rethinking human capital, creativity and urban growth. *Journal of Economic Geography*, **9**(2), 147–67.

The New Climate Economy (2014). Innovation. In *The New Climate Economy, Better Growth, Better Climate* (Chapter 7). Retrieved from https://newclimateeconomy .report/2014/wp-content/uploads/sites/2/2014/08/NCE_Chapter7_Innovation.pdf.

Winden, W., Braun, E., Otgaar, A. and Witte, J. (2012). *The innovation performance of regions: Concepts and cases*. Rotterdam: Erasmus University.

Yigitcanlar, T. and Bulu, M. (2016). Urban knowledge and innovation spaces. *Journal of Urban Technology*, **23**(1), 1–9.

10. Digital nomadism: mobility, millennials and the future of work in the online gig economy

Beverly Yuen Thompson[1]

DIGITAL NOMADS

At the beginning of popular use of the internet, Tsugio and Manners (1997) published their forward-thinking manifesto *Digital Nomad*, a contemplation on the possibility that the internet could provide some workers with the option of remote work. No longer tethered to their office and bosses, they could use their new-found freedom to travel the world, explore exotic locations and work from their laptop, between sips of margarita on the beach. Tsugio and Manners (1997) recognised that some would continue to be tethered to location, such as factory and service workers, while more technologically based jobs could be moved online. They specified that workers whose function is the processing of information in one form or another (graphic artists, writers, designers, software writers and the like) would be most primed for this shift to remote working (Tsugio and Manners, 1997). Decades later, this concept of *digital nomad* was assumed by a subculture of people who identified with this imagined lifestyle. Tsugio and Manners (1997) focused on the liberatory benefits of technology and avoided speaking of the risks for workers.

While proprietary website platforms take the place of face-to-face management of labour, the ways in which workers receive incentives to accept jobs quickly, complete jobs, set their wage and organise their schedule are factored into an algorithm whose calculations are unknown by the workers and the public, yet such algorithms have absolute control over the continued enrolment of the worker in the platform (Kuhn and Maleki, 2017; Schörpf et al., 2017). For example, some platforms require workers to accept a potential job in a certain amount of time, they may have to accept a certain high percentage of jobs, or the platform might encourage workers to continue working with incentives or functions (Kuhn and Maleki, 2017). Platforms also offer the possibility of management surveillance of workers based on registering keystrokes and screenshots (Schörpf et al., 2017, p. 54). Such precarious employment, coupled with a potentially global workforce, and with varying worker-dependencies on the platform (ranging from infrequent workers to full-time workers), can leave workers with considerable financial and lifestyle risk (Graham et al., 2017). Additionally, such work contributes to worker isolation because such work is completed individually online. Platform employers can also engage in potential discrimination based on protected identity classes because there is a lack of oversight and regulation by the government for such cases.

Within the last half-decade, the concept of the digital nomad has been lauded by the press and the business world, as they promote the potential opportunities of flexibility, while obscuring the dark side of the growing economy based on one-time or short-term work, known as gigs, thus the term *gig economy*. As gig work – both location-based, such as Uber, TaskRabbit and Airbnb, and digital platform remote work, such as that provided by Upwork and Amazon's Mechanical Turk – continues to proliferate, so too do the numbers of aspiring workers on a global race to the bottom for wages. Such platforms are notoriously difficult to regulate because governments rely on employment models based on company-based full-time employment, and face challenges on how to regulate such digital piecemeal work (Coyle, 2017).

This chapter is based on empirical data from interviews with 38 individuals encountered at digital nomad conferences in Europe. It is important to understand, within the context of the neoliberal economy, the motivations of workers entering the digital gig economy, and their relative lack of experience working full-time previously before entering this workforce.

THE NEOLIBERAL ECONOMY AND ITS IMPACT ON EMPLOYMENT AND THE URBAN LANDSCAPE

Digital nomads exist in a particular sociological location at the intersections of a post-industrial and speculative finance economy, deregulation and a shift towards a gig economy, replacing the era of job security and *good jobs* (Aronowitz and DiFazio, 2010; Krippner, 2011). In the United States (US) and other Western countries, the post-war era through the 1970s was an era of prosperity and growth, providing even factory workers and other skilled labourers with a middle-class income and benefits. With the technological era growth of the 1990s and onward, the promises of the internet have proved dystopian for workers' rights and benefits. Rather than labour-saving technology sparing the extended working hours of employees, technological algorithms have replaced bosses and such structures have continued to squeeze the employees mercilessly. And yet, the gig economy, as it has come to be called, is heralded in pro-corporate ideological rhetoric as *freeing* workers by allowing them to control certain aspects of their work output, such as their schedule – which is more illusion than reality because those who rely solely on this type of employment must work endlessly, as reported by ethnographers (Ravenelle, 2019; Rosenblat, 2018; Slee, 2017).

This neoliberal and finance-driven economy is having widespread impacts on the look and feel of the gentrifying cities: both rapid development of luxury habitation and extreme inequalities are visible in urban landscapes. And yet, these cities remain highly sought after because the *creative class* (Florida, 2014) of workers is attracted to the white-collar job opportunities that cluster in gentrified, urban hubs, such as the larger cities in the US and across the globe. Moretti (2013) describes this as "the new geography of jobs", driving America into three divergent socio-economic pathways: the brain hubs, the rust belt and the ambivalent middle of the rest of the expansive country, which could go either way.

At a time when inequality in the US rivals that of the divide during the Depression era, all major cities erect construction cranes into the horizon, birthing condominium towers and residential sprawl. And as members of the National Association of Realtors will attest, "while all real estate is local, that does not mean that all property buyers are" (Zhang, 2014, para. 3). In 2013, Chinese investors spent US$12.8 billion in the US residential property market; in 2012, foreign buyers spent over US$68.2 billion in American real estate overall (Zhang, 2014). This is at the same time as someone with full-time employment at the minimum wage cannot pay rent in any of the major cities in the US, all-cash purchasing of residential homes is on the rise, and in California, one in four single-family homes and condominiums is purchased in all-cash transactions (Levin, 2018). Workers are localised, earning the prevailing wages, and yet competing on an international residential housing market with foreign investment finances. And, facing skyrocketing tuition costs in addition to the housing market, it is no wonder that even privileged Western millennials are not necessarily even surviving the neoliberal changes. These millennials live with their parents in increasing numbers and for longer periods of time as they struggle to afford the highly gentrified rents of major urban centres, and even second-tier cities have exploding rent rates far outpacing any rise in income (Desmond, 2017; Hackworth, 2006; Moskowitz, 2017; Rothstein, 2018; Stein, 2019).

The younger generation are highly educated and technologically advanced. Their indulgent parents indoctrinate them with lofty expectations for career options in proportion to their over-education, but such opportunities are few and far between. Having a college degree becomes merely a hoop to jump through without applicable skill development, a taxonomy process of labelling those who can function in the system and in which manner. Thus, for educated and creative millennials who aspire to an international lifestyle, being seduced into the ideologies and possibility of digital nomadism is hard to resist when faced with such a bleak employment landscape. Thus, many may jump first – as the digital nomad ideology of taking a leap of faith encourages – by dropping their entry-level jobs and internships and seeking online employment from internet cafes, in the current study in Chiang Mai, Thailand.

THESE LIQUID TIMES AND DIGITAL NOMADISM

As I have argued elsewhere (Thompson, 2019a), digital nomads are natives to what Zygmunt Bauman (2000) calls *liquid times*: an era marked by uncertainty, diminishing government, rising corporate rule and shifting individual allegiances from country to fractured identities or groups. Nationalist community is "the last relic of the old-time utopias of the good society" (Bauman, 2000, p. 92). For Bauman, such communities in liquid times are temporary, idealised and carnivalesque, but ultimately, are based on consumerism. This consumer society, and the electronic capital that fuels it, moves at the speed of light, with a mobility that can potentially destabilise large swaths of society all of a sudden (Bauman, 2000). This capitalist instability is one more element of the naturally disoriented nature of nomadic life:

> Above all, nomads, struggling to survive in the world of nomads, need to grow used to the state of continuous disorientation, to the travelling along roads of unknown direction and duration, seldom looking beyond the next turn or crossing; they need to concentrate all their attention on that small stretch of road which they need to negotiate before dusk. (Bauman, 2000, p. 209)

The digital nomad is adapted for the crisis of "late capitalist society engaged in a long-term historical process of destroying jobs security, while the virtues of work are ironically and ever more insistently being glorified" (Aronowitz et al., 1998, p. 40). Impoverishment is no longer only for the poor, as middle- and upper-income workers feel the squeeze of the new-order of downsizing, shrinking labour union protection, loss of state welfare and (job-replacing) technologies (Aronowitz et al., 1998). In this world, long-term employment with a family-wage and benefits is dissolved. Tsugio and Manners (1997) specified that workers in web design and computer information would be the most likely to be remote workers. These positions are part of what Richard Florida (2014) has labelled the *creative class*. Florida, and others, such as Sundararajan (2016), touts these industries as primed for the flexibility and creativity that the digital world offers. Duffy (2017) points out that this is a mantra in the fashion industry, where "self-actualisation" is promoted as an outcome of the industry – but in her ethnographic inquiry, Duffy finds that such "aspirational work" is often unpaid and exploitative. In her book *Bait and Switch*, Barbara Ehrenreich (2006) exposes how industries spring up to take advantage of these desperate white-collar employment seekers and how employment fairs and coaching become industries in themselves. Indeed, with their micro-entrepreneurial orientation, digital nomads themselves have been quick to capitalise on conferences for aspirants to the lifestyle with the primary message: dream it and you can achieve it.

Miya Tokumitsu (2015) calls out such industries and tactics in her book *Do What You Love: And Other Lies About Success & Happiness*, and argues that such rhetoric hides the abominable working conditions, especially in the creative industries. Gandini (2016) argues that to be successful, online knowledge workers essentially need to become brands, and the importance of this reputational value to getting work adds new demands on workers. No longer is the curriculum vitae the staple for job applications, but one's rating – which signifies the importance of understanding how platforms and algorithms control workers (Ford, 2016; Gray and Suri, 2019; Kushner, 2013). Overall, contrary to the "freedom" that the business community attaches to the concept of gig work, the work intensification demanded by the low wages and increasing platform fees creates a "new global underclass" that is anything but free (De Stefano, 2016; Gray and Suri, 2019).

WESTERN MILLENNIALS SEEKING EMPLOYMENT IN A NEOLIBERAL ECONOMY

Millennials have attained the highest college completion rates in the US, with 38 per cent of this generation graduating from a four-year degree from a higher education institution (Fry, 2018, p. 54). With this increased rate of higher education achievement, this generation has also endured an astronomical rise in tuition rates and subsequent student loan debt (Geobey, 2018, p. 132). In 2012, the debt-to-asset ratio for the most junior of the age cohorts was over 100 (Fry, 2018, p. 58). The job market they encounter is quickly shifting overhead costs from employer to employee, including healthcare, pension, sick days, security and even office space (Vinodrai, 2018, p. 46). Millennials rely heavily on socio-economic class benefits, if they have access to them, and Worth (2018) and Moos et al. (2018) argue that direct forms of wealth transfer from parents to children perpetuate class-based inequalities in society. In the US, millennials hold only 3 per cent of the national wealth, compared to baby boomers who, at the same age, owned 21 per cent of the wealth (Hoffower, 2019).

While the US is at the forefront of neoliberal cutbacks in social benefits, other countries are following suit, and millennials are especially feeling the squeeze, which I have elaborated on elsewhere (Thompson, 2018; 2019b). Jennifer Rayner (2016), an Australian millennial, wrote *Generation Less*, outlining the loss of status of her generation. Her peers rent much longer than previous generations did, delay purchasing property, if they are able to afford such an expense at all, and have the largest debt loads (especially in relation to student debt) compared with previous generations (Rayner, 2016). Millennials comprise 13.9 per cent of the United Kingdom (UK)'s population, with a 19 per cent concentration in the expensive urban centre of London (Brown et al., 2017, p. 3). Most commonly, millennials are employed in corporate service work, followed by employment in health and educational institutions, but they have endured the largest fall in average earnings (Brown et al., 2017, p. 3). Furthermore, despite porous borders in the European Union, young people encounter a job market that is hard to navigate (Eurostat, 2015).

LOCATION, TOURISM AND THE SEARCH FOR NOMADIC COMMUNITY

Digital nomads select their destination based on where their peers visit. Chiang Mai, Thailand, has become a popular choice because of its affordability and tourist-accommodating orientation. Florida (2014) states that real estate developers are beginning to respond to freelancers' and travellers' needs for flexible work spaces. Thus, digital nomads align with the rise of gentrification, development of co-working spaces and a market targeted towards their Western commodity tastes over the local population, which contributes to the tourist economy and the rising costs for local residents.

Tourism is one of the largest global industries, accounting for nearly 10 per cent of world GDP and 8 per cent of all employment (Urry and Larsen, 2011). Tourism originates in locations unequally: the countries with the highest human development account for most international tourism departures (Urry and Larsen, 2011). Not only is passport strength a primary factor for becoming a digital nomad, but the race of the tourist is also envisioned as white; tourists of colour are not considered (in marketing, for example), and encountering racism during travel is rarely discussed. Caren Kaplan (2002) encourages us to be critical of modern tourism as a practice:

> Moderns value mobility, especially leisure travel, and many of us take travelling for granted. But if travel is central to modernity, then the critique of travel must be a fundamental priority in contemporary critical practices. In this critical approach to deconstructing something that one engages deeply and cares about, the term travel signifies the multiple aspects of an expanded field including transportation and communications technologies, divisions of labor, and representational practice. (p. 32)

While digital nomads like to bracket themselves off from tourism, implying that they are more "global citizens" or serious travellers than the superficial tourists, they are merely a tourist of a longer duration because their relationship to places remains similar: *aesthetic* (Wood, 2005, p. 54). The tourist gazes upon the locals in an objectifying manner, relating to them as fetishised commodities of the local environment (friend, shaman), but the locals also hold an oppositional gaze, or a double-consciousness gaze (Du Bois, 2007; Urry and Larsen, 2011). Although digital nomads consider themselves serious travellers, they continue to bracket themselves off from local communities and consume Western-level accommodations.

RESEARCH METHOD

Ethnographers attempt to understand a social world from the perspective of the participants, how they make sense of their community, experience events and talk about significant concepts (Denzin, 1974; Emerson et al., 1995). I began research in September 2017, when I attended a conference on the topic of nomadism in Europe called "DNX 2017: Digital Nomad Girls Retreat" and then attended the "7in7" Conference in Barcelona in October 2017. Each participant was interviewed either in person or on Skype, for approximately an hour to an hour and a half, about their life trajectory of travel, their general socio-economic social location, their employment history, and their experiences as a digital nomad or an aspirant. The methods and participants are discussed in more detail in Thompson (2018). Participants for this chapter were recruited from both conferences.

PARTICIPANT DEMOGRAPHICS

The 38 participants comprised 33 women and 5 men. They were primarily from Western countries, English speaking, mostly white (n = 28), and primarily of the millennial generation (n = 12 in their 20s, n = 22 in their 30s, and n = 4 in their 40s). Only 6 participants were married (n = 32 were single). Only 1 participant had a child. Only 6 participants had not completed a higher education degree, and half of all of the participants had accumulated student debt.

FINDINGS: DIGITAL NOMADISM AS INDIVIDUALIST ECONOMIC COPING STRATEGY FOR ADVENTUROUS MILLENNIALS

Very few of the participants entered the lifestyle of digital nomadism from a place of employment security, or a lucrative salary, from which to securely afford travel. They came from Western countries and had privileges based on citizenship, social class of parents, social capital, college education and race. Half of the participants had student debt, such as Elizabeth from Canada: "So much. I have about [CA]$40,000 of student debt. Canadian. Yeah. For five years. I did an extra year just to get even more in debt. I'm in deferment. I make no on-paper income, [so] I don't have to pay anything back."

For some participants, from Commonwealth nations such as Canada and Australia, student debt becomes a government tax taken as a very small percentage of income – when there is income. For the Americans, they had to begin repayment within zero to six months of graduation, regardless of their employment status, with payments often as large as rent. Anna, a multi-racial American, spent her childhood in Pakistan until her parents' divorce. She attended an elite, expensive college:

> Bryn Mawr was around [US]$50,000 a year. I could not have got to Bryn Mawr without financial aid. And Bryn Mawr at the time was "need blind": if you can get in, they'll help you pay.... It was [US]$12,000 or [US]$15,000 of debt. For work, I was cobbling it together. I never had a full-time job. When I graduated, I was tutoring French, I was babysitting and I was on the streets of New York City handing out coupons and ice cream samples...

Anna was intelligent, spoke multiple languages that she taught for extra cash, and had professional parents. She was not able to land a fulfilling full-time job and gave in to her wanderlust, hoping to find more opportunities outside of the US:

> It was too cold in the US. It was too expensive. I couldn't afford to rent in New York, while I was recovering from an injury and not earning money. I could not stay at my mom's house because her husband's a jerk. And I wasn't supposed to be stressed out because of the brain injury. So, I ended up meeting back up with some entrepreneurs at a retreat for a month in Thailand. I spent a time with really interesting, intelligent people.

For young Americans, having medical issues, or medical debts, can be another contributing factor towards financial precarity. Living in the most expensive American city, while suffering health complications and finding no full-time work opportunities, made it an easier path to explore Thailand and join the communal attempt at figuring out some kind of remote work option. Aspiring nomads were told about opportunities that would provide "passive income" and could make them rich, while they took surfing lessons and drank cocktails. Elizabeth told me about how she first left Canada, found precarious work through a friend and then fell in with the nomad crowd at wi-fi cafés in Thailand. She takes up *drop-shipping*, a common entry-level remote work that nomads encourage each other to enter into in their workshops and online literature. Drop-shipping is when a person or company sets up a website to sell various products, which are then actually distributed by the manufacturer, not the website owner. Usually selling various low-end knick-knacks, such aspiring sales people soon realise how difficult it is to drive traffic to their new website and they often receive very little income from such endeavours.

> I had heard of Tim Ferriss but later came to know his [book] *The 4-hour Workweek*. I didn't know about drop-shipping. I didn't know that you could be location independent. I didn't know any of these terms or any of the people that talk about it. I moved to Thailand to teach English. I didn't really have direction. I was living with my boyfriend and we broke up. I moved out. I ran into a friend who said, "I'm teaching English in Thailand". I knew someone who moved to Thailand a year before to teach English and her Instagram was unreal. Her life looked so amazing. I was like, "I've got nothing to live for anymore. How do you find a job that's not in your country? I can't even find a job in my own city and I live here." After some back and forth, [my friend in Thailand] said, "My school is hiring. I'll see if I can get you a job." He later said, "Hey, you got a job. School starts on the 15th. You have to be there on the 1st." [I thought] It's April 16th today and you want me to be there in how many days? Are you fucking kidding? Luckily, I have a quasi-supportive family. I was like, "Look. I wanna go teach English in Thailand. I have to leave in 2 weeks. I need money. My birthday is coming up. Make this my birthday present. I'm going and you have to help pay for this." So, that's another side thing about how not everyone can just dive into this. I was very lucky and very fortunate that everyone chipped in a couple [of] hundred dollars and got me a ticket.

While the nomads had difficulty landing full-time employment, they had a great deal of social capital and class standing via their parents and their college education, from which they could wrest enough small gains. Others were rejecting the path laid out for them by their social class standing and familial expectations by taking an alternative route, which left their family and friends flabbergasted. Jenny, the founder of Digital Nomad Girls, and her partner Simon, had both earned PhDs in chemistry from the UK, but decided after those years in the laboratory that they did not wish to work in industry. While meeting digital nomads in Thailand, the couple went through the gamut of possible remote jobs. Jenny started her small retreat business while Simon supported them by writing chemistry abstracts for academic journal publications. Simon told me about their start in the lifestyle:

> We got caught up in this drop-shipping phenomenon. It's an online store where you sell other people's stuff. It's not easy to get people to buy from your store. It was one of these things we'd heard about through podcasts. It's not a scam exactly, but people talk about it like it's a dream job with which you can make passive income. Jenny realised she wasn't interested in it. I tinkered with it for a couple of months, part-time, and got bored. There was a whole scene of people trying to sell you stuff in Chiang Mai. At the same time, we met loads of people struggling, loads of people doing well. It is a lovely place to be. It has Western comforts. There is a lot of local stuff. About the only thing they don't have is a little bit of coastline to go for a swim. But, it's a really good place to start out. It has become a bit of a cliché now, but I would still recommend it to people, particularly to those who don't have

that much money when starting out. You can live cheaply. You can have a very nice comfortable lifestyle on very little.

These stories of downward mobility in one's home country, lack of interest in pursuing the careers foisted upon them by parents and college professors, and a desire for adventure and meaning in life, led these nomads to travel and find a way to continue the path. A small number of the nomads had previously been employed full-time in high-salary occupations, such as computer programming. Lewis was a white male programmer from the UK: "I've been able to live mostly off the income for my apps. Finding work was mostly relationship building. I prefer to go to co-working spaces because I like working in an office."

Nick and Lewis were friends, and both had previously worked full-time jobs with salaries in the six-figure range for several years before being laid off and finding still-lucrative consulting work in their field. But they never matched their full-time salary and benefits with their remote consulting work. Nick said:

> When I graduated from Georgia Tech, I moved to a full-time job at a company for six months to a year. The cost of living in LA [Los Angeles] is a lot more than in Atlanta, Georgia, so I had to find higher-paying employment. It was a bit of a dark period. I left that job and did some freelance work, software development. Then, I got a job at a company that was owned by AOL [America Online Inc.], my last full-time job before I became a digital nomad. It was a six-figure salary, health benefits, 401K, and all the corporate bits to the nines. I got that right at the beginning of 2007, so I was there a little under two and a half years. AOL has this annual purge where they equalise the books. So, I'm on holiday with Kit, my partner, in Norway. They fire everyone in the company and shut it down. That is what kick-started us on becoming digital nomads. This was back in mid-2009. I get handed this massive severance package. We're like, "This will only last 2 months in Los Angeles OR…" I think we did a full round-the-world trip and I think the money lasted almost six months.

While a few of the men had training in lucrative fields such as computer programming, only two of the women were so established in the same profession. Hannah Wei, a successful computer programmer, took leave from university while she worked for a company and has sold some of her startups for enough financial security to provide herself with savings and health care. She also uses her technological work to help local community organisations in Asia and Africa. She was a speaker at the 7in7 conference where she challenged the digital nomads on their privilege and their relationship to local communities. Hannah Wei told me a bit of her story:

> My startup was called Vizualize.me. It's a web app that takes your résumé and turns it into a one-page infographic. I worked on that with two other people, maybe a few more. I used up all my savings, the typical story with startup life. I actually took time

off school to work on this as well, so I didn't go back for my fourth year. I finished my degree later. We eventually sold the business. And then that spun off another startup. The narrative around "inspiration", "you need to hustle", there's a very dark side to that, and I see it in the startup environment. It's a very toxic thing. And you really don't take into consideration what resources people have access to, how much leverage that their race and their education level and how much debt they have would give them. I think some people just don't have that leverage to afford to take one year and "be an entrepreneur". Like, for me, that's not entrepreneurship because they're not creating new jobs, they're not building a community outside their own brand, and they're not contributing to society in any form. They're not contributing to taxes wherever they're living. So that turned me off. I started digging into who are some of the personalities in digital nomadism, who understand that this is not sustainable, who understand that the work that they do has an impact on people in the digital nomad space, and also the communities that they live in.

For the most part, the women who had started their micro-businesses were selling the digital nomad lifestyle to other aspiring (primarily women) nomads, but rarely making above a subsistence level of money. Jenny began attempting to monetise her Digital Nomad Girls Facebook group through organising retreats after she and her partner Simon attempted a series of popular remote work gigs before abandoning each in turn. She told me:

We moved to Chiang Mai in June 2015. The idea was kind of sparked more in the co-working spaces of Chiang Mai because I met my friend Marta there. We were usually the only girls. At meet-ups, it was very much still dominated by guys. So, I decided to start a Facebook group and it went from there. It was just a way of meeting people. I think we were up to 1,000 quickly. We are at nearly 14,000 girls now. In the last year, we added about 10,000 people to the group.

However, Jenny had organised fewer than one retreat each year, bringing 15 women together, which did not equate to financial earnings after covering the expenses of the events. Her partner Simon supported both of them through his gig of writing chemistry abstracts for academic journals, which hardly provided a living wage. As Simon expressed, he hoped that in the future Jenny's retreat business could provide more of a stable income. In the year since I attended the retreat, a subsequent one has yet to be organised. Other attendees at such retreats also spoke of starting similar businesses that make a survival income off other nomadic aspirants' buying their digital products.

DISCUSSION AND CONCLUSION

Millennials face dire employment opportunities under the current neoliberal trajectory, which Bauman (2000) refers to as liquid times. According to Bauman (2000), these liquid times represent a trajectory of dismantling of master-narratives and institutions, diminishing of governments, the continuing empowerment of corporate domination and a shift away from an identity aligned with the nation-state. The creative classes of workers (Florida, 2014) are uniquely positioned to have their job opportunities ravaged by Silicon Valley corporate platforms into piecemeal gigs, as corporations move away from full-time employment with benefits. Everything from healthcare to office space must be supplied by the worker. The millennial generation is most primed to be subsumed by this gig economy as they encounter their first entry-level employment, or piece together various part-time gigs. For an adventurous few, they realise their Western metropolitan areas are increasingly unaffordable, as global capital is further invested in real estate in major global cities (Zurkin, 2014). Thus, digital nomads use the opportunity of working remotely, or finding remote work after they decide upon this lifestyle, and move to a country such as Thailand, where its luxuries become affordable to a downwardly mobile Westerner. In other words, the digital nomad is adapted for the crisis of "late capitalist society engaged in a long-term historical process of destroying jobs security, while the virtues of work are ironically and ever more insistently being glorified" (Aronowitz et al., 1998, p. 40). The digital nomad mantra, akin to their patron saint Tim Ferriss (2009), is one that promotes willpower as singly determining ("visualise it and it will happen") and a leisurely existence maintained by "passive income" ("make money while you sleep").

Unlike the stories sold at digital nomad conventions, stories of the participants reflected downwardly mobile trajectories of otherwise highly educated and high social-capital holding millennials, raised with indulgent aspirations of grandeur instilled in them by indulgent parents. Half of them had significant student debt (especially the Americans), with the other half often having access to government grants in Commonwealth countries. The participants were of a middle-class background with the expectation of a college education and employment in their chosen field, and did not think deeply about taking out such debt because it had become normalised and expected. Their debt often was not as bad as some of their peers, they self-consoled. Few had had full-time employment with benefits before embarking on their lifestyle adventure, except for the few who had a background in computer technical support and computer programming. The employment division was especially stark when examining gender: the men of the sample previously had high-income computer jobs, compared with the women, who were getting their start selling Facebook ads, being a virtual assistant or working in web design. Many of the participants jumped into the lifestyle before finding remote work. They used their social networks in tourist hot-spots to learn about get-rich-quick schemes such as drop-shipping. Their conferences covered lifestyle topics such as finding work, avoiding loneliness, meeting other singles on the road or the challenges of travelling as a family. The answers were always optimistic.

However, the challenges of constant movement, irregular schedules, lack of socialising and strains on interpersonal relationships seemed to cut the lifestyle short after a few years. For example, workers may need to maintain business hours in the US or Europe, but their location in Thailand may make this inconvenient for keeping a regular daytime work schedule. The challenge of constantly planning new travel agendas, leaving and entering a country to renew a tourist visa, or simply not having access to clothes-washing machines in Airbnb temporary homes takes an extra layer of planning and work.

Finally, from this qualitative study, after getting some post-research updates from participants, it seems that a large percentage do settle somewhere, often back in their original hometown or previous residence in their home country before they embarked on this lifestyle. Overall, the work they were able to find was the essence of gig work, piecemeal employment that is difficult to maintain and make a living from. Those working full-time for a company were the most successful in the short term, but such jobs rarely lasted more than a few years. Most were not developing marketable professional skills that would lead to full-time employment with a company and were overall representing a downward mobility, rather than an upward professional trajectory. The increasingly online digital gig opportunities offered to contemporary workers is rising at the same time that the expectation of long-term, full-time employment with benefits with a singular company is disappearing. The stories of young digital nomads point to a future of work that is increasingly insecure. In order to soften such trajectories, workers coming together at digital nomad summits should use their collectivity in order to fight for further worker protection on such platforms and through governmental regulations. While being a digital nomad may provide opportunities to travel and engage in serious leisure, it does not ensure a future of work security.

NOTE

1. ORCID iD: 0000-0001-7375-2789, Department of Sociology, Siena College, bthompson@siena.edu.

REFERENCES

Aronowitz, S. and DiFazio, W. (2010). *The jobless future* (2nd edn). Minneapolis, MN: University of Minnesota Press.
Aronowitz, S., Esposito, D., DiFazio, W. and Yard, M. (1998). The post-work manifesto. In S. Aronowitz and J. Cutler (eds), *Post-work* (pp. 31–67). New York, NY: Routledge.
Bauman, Z. (2000). *Liquid modernity*. New York, NY: Polity.
Brown, J., Apostolova, V., Barton, C., Bolton, P., Dempsey, N., Harari, D., Hawkins, O. et al. (2017). *Millennials* (House of Commons Library Briefing Paper No.

CBP7946, 11 April). Retrieved from UK Parliament website: researchbriefings.files .parliament.uk/documents/CBP-7946/CBP-7946.pdf.

Coyle, D. (2017). Precarious and productive work in the digital economy. *National Institute Economic Review*, **240**(1), R5–R14.

Denzin, N.K. (1974). The methodological implications of symbolic interactionism for the study of deviance. *The British Journal of Sociology*, **25**(3), 269–82.

Desmond, M. (2017). *Evicted: Poverty and profit in the American city*. New York, NY: Broadway Books.

De Stefano, V. (2016). The rise of the "just-in-time workforce": On-demand work, crowd work and labour protection in the "gig-economy" (Conditions of Work and Employment Series No. 71). Retrieved from International Labour Office website: https://www.ilo.org/wcmsp5/groups/public/---ed_protect/---protrav/---travail/ documents/publication/wcms_443267.pdf.

Du Bois, W.E.B. (2007). *The souls of black folk*. New York, NY: Oxford University Press.

Duffy, E.B. (2017). *(Not) getting paid to do what you love: Gender, social media, and aspirational work*. New Haven, CT: Yale University Press.

Ehrenreich, B. (2006). *Bait and switch: The (futile) pursuit of the American dream*. New York, NY: Holt Paperbacks.

Emerson, R.M., Fretz, R.I. and Shaw, L.L. (1995). *Writing ethnographic fieldnotes*. Chicago, IL: University of Chicago Press.

Eurostat (2015). *Being young in Europe today*. Luxembourg: Publications Office of the European Union.

Ferriss, T. (2009). *The 4-hour workweek: Escape 9–5, live anywhere, and join the new rich*. New York, NY: Harmony Books.

Florida, R. (2014). *The rise of the creative class*. New York, NY: Basic Books.

Ford, M. (2016). *Rise of the robots: Technology and the threat of a jobless future*. New York, NY: Basic Books.

Fry, R. (2018). Young adult household economic well-being comparing millennials to earlier generations in the United States. In M. Moos, D. Pfeiffer and T. Vinodrai (eds), *The millennial city: Trends, implications, and prospects for urban planning and policy* (pp. 53–66). New York, NY: Routledge.

Gandini, A. (2016). *The reputation economy: Understanding knowledge work in digital society*. London: Macmillan Publishers.

Geobey, S. (2018). Planning for the sharing economy. In M. Moos, D. Pfeiffer and T. Vinodrai (eds), *The millennial city: Trends, implications, and prospects for urban planning and policy* (pp. 128–49). New York, NY: Routledge.

Graham, M., Hjorth, I. and Lehdonvirta, V. (2017). Digital labour and development: Impacts of global digital labour platforms and the gig economy on worker livelihoods. *Transfer*, **23**(2), 135–62.

Gray, M. and Suri, S. (2019). *Ghost work: How to stop Silicon Valley from building a new global underclass*. New York, NY: Houghton Mifflin Harcourt.

Hackworth, J. (2006). *The neoliberal city: Governance, ideology, and development in American urbanism*. Ithaca, NY: Cornell University Press.

Hoffower, H. (2019). Millennials only hold 3% of total US wealth, and that's a shockingly small sliver of what baby boomers had at their age. *Business Insider*, 5 December. Retrieved from: https://www.businessinsider.com/millennials-less -wealth-net-worth-compared-to-boomers-2019-12.

Kaplan, C. (2002). Transporting the subject: Technologies of mobility and location in the era of globalization. *PMLA*, **117**(1), 32–42.

Krippner, G.R. (2011). *Capitalizing on crisis: The political origins of the rise of finance*. Cambridge, MA: Harvard University Press.

Kuhn, K.M. and Maleki, A. (2017). Micro-entrepreneurs, dependent contractors, and instaserfs: Understanding online labor platform workforces. *Academy of Management Perspectives*, **31**(3), 183–200.

Kushner, S. (2013). The freelance translation machine: Algorithmic culture and the invisible industry. *New Media & Society*, **15**(8), 1241–58.

Levin, M. (2018). Data dig: Are foreign investors driving up real estate in your California neighborhood? *Cal Matters*, 7 March. Retrieved from https://calmatters.org/housing/2018/03/data-dig-are-foreign-investors-driving-up-real-estate-in-your-california-neighborhood/.

Moos, M., Pfeiffer, D. and Vinodrai, T. (eds) (2018). *The millennial city: Trends, implications, and prospects for urban planning and policy*. New York, NY: Routledge.

Moretti, E. (2013). *The new geography of jobs*. New York, NY: Mariner Books.

Moskowitz, P. (2017). *How to kill a city: Gentrification*. Lebanon, IN: Bold Type Books.

Ravenelle, A.J. (2019). *Hustle and gig: Struggling and surviving in the sharing economy*. Oakland, CA: University of California Press.

Rayner, J. (2016). *Generation less: How Australia is cheating the young*. Carlton, VIC: Redback Quarterly.

Rosenblat, A. (2018). *Uberland: How algorithms are rewriting the rules of work*. Oakland, CA: University of California Press.

Rothstein, R. (2018). *The color of law: A forgotten history of how our government segregated America*. New York, NY: Liveright.

Schörpf, P., Flecker, J., Schönauer, A. and Eichmann, H. (2017). Triangular love–hate: Management and control in creative crowdworking. *New Technology, Work and Employment*, **32**(1), 43–58.

Slee, T. (2017). *What's yours is mine: Against the sharing economy*. New York, NY: OR Books.

Stein, S. (2019). *Capital city: Gentrification and the real estate state*. New York, NY: Verso.

Sundararajan, A. (2016). *The sharing economy: The end of employment and the rise of crowd-based capitalism*. Cambridge, MA: The MIT Press.

Thompson, B.Y. (2018). Digital nomads: Employment in the online gig economy. *Glocalism: A Journal of Culture, Politics, and Innovation*, 1.

Thompson, B.Y. (2019a). "I get my lovin' on the run": Digital nomads, constant travel, and nurturing romantic relationships. In C. Nash and A. Gorman-Murray (eds), *The geographies of digital sexuality* (pp. 69–90). Singapore: Palgrave Macmillan.

Thompson, B.Y. (2019b). The digital nomad lifestyle: (Remote) work/leisure balance, privilege, and constructed community. *International Journal of the Sociology of Leisure*, **2**, 27–42.

Tokumitsu, M. (2015). *Do what you love: And other lies about success & happiness*. New York, NY: Regan Arts.

Tsugio, M. and Manners, D. (1997). *Digital nomad*. New York, NY: John Wiley & Sons.

Urry, J. and Larsen, J. (2011). *The tourist gaze 3.0*. London: SAGE Publications.

Vinodrai, T. (2018). Planning for "cool": Millennials and the innovation economy of cities. In M. Moos, D. Pfeiffer and T. Vinodrai (eds), *The millennial city: Trends, implications, and prospects for urban planning and policy* (pp. 45–51). New York, NY: Routledge.

Wood, M. (2005). Nomad aesthetics and the global knowledge economy. *Journal of Critical Postmodern Organization Science*, **3**(3/4), A50–64.

Worth, N. (2018). The privilege of a parental safety net: Millennials and the intergenerational transfer of wealth and resources. In M. Moos, D. Pfeiffer, and T. Vinodrai (eds), *The millennial city: Trends, implications, and prospects for urban planning and policy* (pp. 110–27). New York, NY: Routledge.

Zhang, A. (2014). Chinese are No. 1 buyers of US residential property. *China Daily*, 10 July. Retrieved from http://usa.chinadaily.com.cn/world/2014-07/10/content _17698349.htm.

Zurkin, S. (2014). *Loft living: Culture and capital in urban change*. New Brunswick, NJ: Rutgers University Press.

11. Playing with TikTok: algorithmic culture and the future of creative work

Natalie Collie[1] and Caroline Wilson-Barnao[2]

INTRODUCTION

Before the young American rapper Lil Nas X signed with Columbia Records in early 2019, his country-trap single "Old Town Road" went viral with the help of TikTok, a popular video-sharing app owned by Chinese social media giant ByteDance, with over 500 million active monthly users worldwide. Lil Nas X promoted the meme for months after uploading it to the platform, where it eventually became the soundtrack for the #yeehaw challenge, a game in which users filmed themselves transforming into cowboys and cowgirls after drinking "yee yee juice", while lip-syncing "Old Town Road" (Strapagiel, 2019). The videos themselves[3] are amateur, often cringeworthy, sometimes hilarious and highly addictive: typical of the playful meme-driven video-sharing platform. Lil Nas X's use of TikTok for promotional purposes is also an increasingly important, if not typical, story of professional creative practice: #yeehaw videos, almost all of which sample "Old Town Road", have been viewed over 67 million times (Chow, 2019) and the rapper credits the platform with the original success of the single (Strapagiel, 2019). After Lil Nas X signed with Columbia Records, his song and its famous remix with Billy Ray Cyrus broke numerous records, peaking for 19 consecutive weeks on the Billboard Hot 100 chart in the US.

A further intriguing element in this story of platform-driven professional success is the fact that it was built on Lil Nas X's previous experience creating memes and going viral on Twitter as a teen tweetdecker.[4] Some in the industry speculate that his decision to originally list the single on SoundCloud and iTunes as a country record was also strategic. For example, Danny Kang, co-manager of another viral country musician, said that Lil Nas X intended to gain "a little bit of traction on their country charts, and…manipulate the algorithm…versus trying to go the rap format to compete with the most popular songs in the world" (Leight, 2019, para. 8).[5] Lil Nas X's story provides a striking illustration of the transformations of creativity, play and work underway in an *algorithmic* culture – one increasingly organised by the imperatives of big data and associated analytic capability (Striphas, 2015).

In this chapter, we use an analysis of TikTok to trace the production of creative work *and* play organised by contemporary media platforms. While platforms such as TikTok, YouTube and Instagram are transforming the work of creative professionals across a host of industries, they are also remaking vernacular culture and other dimensions of human experience, including play and creativity, into a kind of immaterial and unpaid *digital labour* (Terranova, 2000). These seemingly democratised systems of cultural production and distribution allow anyone, even children, to produce, curate and share creative content. These systems function to embed creative, playful practices and their value within the data-driven and commercial logic of the digital economy, including ubiquitous forms of ever more granular surveillance and data capture. Culture is increasingly dominated by an imperative on all of us to be creative – to express ourselves, to build a personal brand and to curate our lives – yet this identity work is outsourced to the mediation of digital platforms. And this imperative has become embedded in our social identity, relationships and the logistics of our everyday lives. As a consequence, the conventional distinctions between work and play, between professional and amateur, between active production and passive consumption, and between producers and audiences have been fundamentally eroded, transforming the nature of creativity and creative work in the process.

We argue that TikTok needs to be understood as a particular *architecture of digital labour* (Postigo, 2016), designed to efficiently harness the playful creativity of its users in ways that not only build engagement but that also, in turn, shape the nature and the value of that play. What matters in such a system is not the specific content or its meaning – as such – but rather its capacity to incite interest, be shared and go viral. This, we suggest, also has significant implications for future creative-work practices and how they are valued.

PLAYING WITH ALGORITHMS

Lil Nas X's success at *gaming* or *playing* TikTok and other social media platforms illustrates the importance of creatives understanding how platform culture works, in particular the operations of content-recommendation algorithms. Lil Nas X's story also exemplifies more broadly the notion of *algorithmic culture* developed by Striphas (2015) and others to characterise the "enfolding of human thought, conduct, organization and expression into the logic of big data and large-scale computation, a move that alters how the category *culture* has long been practiced, experienced and understood" (p. 396).

Indeed, an analysis of TikTok's *socio-technical architecture* (van Dijck, 2013) highlights the extent to which digital media platforms not only affect professional creativity, but also, increasingly, have an impact on everyday play and children's culture, transforming our understanding of both creative work *and* creative play in the process. For example, growing up during the 1980s, we enjoyed re-enacting our favourite music videos, singing the lyrics and dancing at home with our friends. While we might have dreamt of fame or talked about charging admission from parents or neighbours, this game remained private, ephemeral and unmediated. And no one profited from our creative labour. The contemporary mediated performances of young people on TikTok are not dissimilar to the ones that we made in the 1980s. However, they are now organised and made public by a social media platform. The TikTok app both extends and limits the creative process. It presents children with a series of options from which they can choose to enhance their performance, but it also limits their creative labour by pre-empting a series of practices and performances of identity – sometimes disturbingly sexualised – with which they are intended to engage. The commenting and feedback system produces interactions that give a sense of intimacy, connection and attention. TikTok is a digital stage where children perform *virtually* in front of their friends and others, demonstrating their musical talent, ideas and creative mastery of the platform.

We argue, then, that creative engagement or *playing* with algorithms can transform the experience of users and questions of work, creativity and value. In Lil Nas X's case, savvy use of meme culture allowed him to launch a commercial, mainstream career in a way that might have been almost impossible only a few years ago. Indeed, his success seemed so unexpected that he was suspected of being an *industry plant* (Reily, 2019) – the managed product of a calculated marketing strategy by industry operators – with the suggestion of inauthenticity and questionable talent that the accusation implies. In the case of re-enacting music videos, children's playfulness is now mediated and transformed by a platform organised by this very same logic of spectacle and share-ability, one that, in their instance, also invokes questions of privacy and surveillance in the context of under-age users. Our analysis of the socio-technical architecture of TikTok – using Light et al.'s (2018) walkthrough method developed specifically for the analysis of apps – is designed to trace this *datafication* of creativity and its impact on cultural production and consumption.

ANALYSING TIKTOK

Unsurprisingly, given that it was only launched in 2017, there is little scholarship yet published about TikTok. Our study builds on a small body of literature focused on its predecessor, Musical.ly.[6] Literat and Kligler-Vilenchik (2019), for example, examine the use of Musical.ly by young people during the 2016 presidential elections in the US. Their research suggests the importance of such platforms for young people's political expression via *shared symbolic resources*, the use of hashtags in particular. This question of shared symbolic resources also informs Rettberg's (2017) focus on the use of hand signs in lip-sync videos, arguing for their equivalence to the use of emojis in text-based forms of communication. Of most relevance to our research, however, is Simsek et al.'s (2018) use of the walkthrough method (Light et al., 2018) to assist in a study of the micro-celebrity practices afforded by the platform and its everyday social use by young girls. We revisit and extend Simsek et al.'s (2018) use of this method with three specific questions in mind. First, what is TikTok's socio-economic and cultural *environment of expected use* (Light et al., 2018, p. 889) informing users' experience of the technology, including its operating model, governance and intended purpose? Second, how does the app's technical architecture configure user activity, in particular their creativity, connection and play? And, third, how does this differ from conventional understandings of creative work and play? Our overall aim is to trace the extent to which TikTok's socio-technical apparatus anticipates a certain kind of user – a user exemplified by both Lil Nas X *and* contemporary tweens and teens – whose creative work is, in turn, shaped by the app's affordances and algorithmic processes.

Understanding TikTok's *Environment of Expected Use*

In 2017, ByteDance, a Chinese technology company that owns and operates a range of digital media platforms, acquired Musical.ly for US$1 billion (BBC, 2019). Already popular in its own right with teenagers in the West, Musical.ly, which was itself a Chinese start-up, had around 60 million active users at the time it was purchased (BBC, 2019). ByteDance then merged the popular lip-syncing app with Douyin, a similar Chinese-based platform, to create TikTok.

While similar in terms of their interfaces and broad appeal with teens and tweens, TikTok and Douyin have different cultural audiences. Douyin is supplied through Chinese app stores to a distinctly Chinese market, whereas non-Chinese audiences are able to access TikTok directly through the Apple store and Google Play (Verberg, 2019). ByteDance negotiates different censorship expectations by operating the two apps from different servers and deploying localised approaches to content moderation and regulatory compliance. This allows Douyin to comply with Chinese government sensitivities at home, while allowing TikTok – in theory, at least – to operate relatively free of these constraints in other jurisdictions (Verberg, 2019).

Nevertheless, TikTok is under increased scrutiny with regard to questions of censorship, child safety and privacy. In February 2019, for example,

ByteDance was fined US$5.7 billion for breaching American privacy law by collecting personal information on users under 13 (Matsakis, 2019a). In response, TikTok has launched a range of measures including stronger age restrictions, enhanced screen-time management, a restricted mode and a video tutorial series on privacy settings and digital wellbeing. More recently, the app was briefly banned in India, its fastest growing market, with legislators arguing that the platform was exposing teens to paedophilia and pornography (Parkin, 2019).

There are also growing concerns that TikTok is censoring political and other kinds of messages in the course of its content moderation. Many fear that civil unrest in Hong Kong, for example, is being actively censored by the platform's mix of human and algorithmic moderators (Harwell and Romm, 2019). Also, a recent leak of internal moderation guidelines (Hern, 2019b) suggests a range of topics that have been actively censored by the platform, including Tiananmen Square, Tibetan independence and pro-LGBT (lesbian, gay, bisexual, transgender) content. The role of algorithms in this process is crucial: if the content is not deleted, its circulation can still be suppressed by the content algorithm, making it difficult for a user to determine if their content has been moderated or has simply failed to go viral (Hern, 2019b). Broader concerns about data sharing with the Chinese government led the United States government to announce a national security review of TikTok in late 2019 (Nicas et al., 2019).

Nevertheless, TikTok is one of the world's fastest-growing social media apps, with over one billion downloads in total globally in early 2019 and boasting the fourth highest download rate of any non-game app in 2018 (Hamilton, 2019). Over 40 per cent of its more than 500 million users are young, between the ages of 16 and 24 (this figure underestimates the many under-age users, of course) and India constitutes its fastest-growing market (Beer, 2019). In mid-2019, ByteDance was valued at more than US$75 billion, making it the most valuable start-up in the world at the time (Tolentino, 2019).

At the time of our analysis, in late 2019, TikTok includes some advertising, but is still under-monetised and under-professionalised relative to many other more established platforms. The platform has introduced a mix of traditional advertising (launch-screen and in-feed video ads) and native advertising, sponsored hashtag challenges in particular. TikTok has also started to integrate URLs on individual profiles that link the user to e-commerce sites (Singth, 2019). However, the platform is clearly focused on growing user engagement at this stage of its development. An immense advertising budget – more than US$1 billion in 2018 alone (Lorenz, 2019) – is targeting users on other platforms such as SnapChat, Instagram and Facebook (Hern, 2019a). TikTok is also paying influencers from other platforms such as YouTube to produce TikTok videos, in order to further build its audience (Lorenz, 2019).

Unlike YouTube, there is currently no in-app means for influencers to make money from the number of people who view their content or any associated advertising. TikTok's Virtual Items system, however, allows viewers to send virtual gifts or coins using Apple or Google payment systems during livestreams. TikTok's Diamond system is then used as a way to acknowledge the popularity of a content curator and forms an in-house virtual credit system based on the number of gifts that a particular creator receives. Individual influencers can then cash out their Diamonds via Paypal with a total weekly withdrawal of US$1000 (Arch, 2019). Content producers can also make money on TikTok via individual sponsorship deals with brands they broker themselves and some creators try to shift their fanbase to other social media platforms.

This lack of monetisation options for influencers cannot last, of course, if TikTok is to maintain content-producer engagement, and seems particularly noteworthy given the unusually high ratio of active production to passive consumption of content currently enjoyed on the platform (Beer, 2019). The platform has given a clear signal of its interest in facilitating greater marketing activity, with the recent launch of TikTok's Creator Marketplace, a service designed to broker relationships between brands and content creators for marketing campaigns.

And yet, this relative lack of professionalised content production, as distinct from YouTube, and slick influencer culture, in contrast with Instagram, seems central to TikTok's current appeal and branding. TikTok (n.d.) describes the platform's mission as "inspiring creativity and bringing joy" (para. 1) and asserts its aim as "to capture and present the world's creativity, knowledge, and precious life moments, directly from the mobile phone. TikTok enables everyone to be a creator, and encourages users to share their passion and creative expression through their videos" (para. 2). This discourse of positive creativity, entertainment and digital playfulness is also used to support TikTok's heavy-handed approach to moderating political or contentious content, including the announcement in October 2019 that the platform would be banning all political, issue and advocacy advertising of any kind (Perez, 2019): "it wants to be known as a place for creative expression, and one that creates a 'positive, refreshing environment' that inspires that creativity" (para. 5).

The "engineered playfulness and performativity" (Savic and Albury, 2019, para. 8) of the platform is also crucial to its capacity to target very young users. The emphasis on creativity, joy and playfulness – rather than social networking, as such – helps to "alleviate parental concerns associated with children's use of social media" (Savic and Albury, 2019, para. 7). One of the reasons that TikTok's predecessor Musical.ly was so under-researched, perhaps, relates to the perception that the app was no more than a high-tech children's "toy" and merely "child's play".

Central to this vision for TikTok and its users is the role of the technology itself, with ByteDance (n.d.) stating:

> Technologies such as virtual reality enabled various content creation tools such as hair-color-change effect and AR [augmented reality] stickers for photo and video making. Our technologies also simplify video editing and suggest music, making it easier than ever to craft creative videos and share them with people of the same interests. (para. 3)

ByteDance positions itself in the marketplace as a company that "deals in AI [artificial intelligence] development such as machine learning and original algorithms" (BeautyTech.jp, 2018, para. 9). The quality of user experience is predicated on this technology, with ByteDance (n.d.) further explaining:

> There is no need to input individual preferences or specify topics of interest when using ByteDance content platforms. With smart recognition technology, our platforms learn about each user's interests and preferences through interactions with the content – including likes, dislikes, comments, report and follow. The more users interact with content, the more relevant and engaging the experience becomes. (para. 2)

When looking to the application of ByteDance technologies, it is noteworthy that Douyin is already operating in-video search, which allows users to search a face to find other videos with that particular individual, and now has advanced e-commerce integration. This makes it possible for users to search videos for clothing and accommodation, for example, and purchase products directly within the app (Niewenhuis, 2019). These strong image-recognition technologies are likely to be used by TikTok in the foreseeable future, particularly as the platform has already confirmed that it is currently testing an in-app shopping affordance with a small number of influencers (Wodinsky, 2019).

This emphasis on the relationship between technology and creative, entertaining content – as opposed to social networking, news and "political" content – threaded throughout TikTok's marketing is also embedded in the technical infrastructure of the platform itself, which will be examined in more detail in the next section of this chapter. Having considered TikTok's operating model, governance and purpose, we now turn a critical eye to the technical part of the walkthrough, where we navigate TikTok's infrastructure from the point of view of a user at the stages of registration and entry, everyday use, and exit and suspension (Light et al., 2018).

Experiencing TikTok's Technical Infrastructure

A user can join TikTok by logging on to the website and by texting themselves a link to download the app. Alternatively, they can download the app from the Apple App Store, Google Play or Amazon Appstore. In the privacy and safety sections, there are options to control who can find the user, post comments, perform duets, react to videos, message the user, view their videos, comment and download content. There is also an option to block specific people.

When the user first registers for the app through a mobile phone, they are asked to establish a private or public account. It is possible to then sign up by entering a phone number or email address, or to join by using Facebook, Google or Twitter data. By entering birthday details and a password, the user can update their profile with a photo, video and biography, and sync it with their Instagram and YouTube accounts.

By allowing the app to access the microphone and camera on their smartphone, the user is instantly enabled as a cultural creator to create a video and select from around 100 different trending, new, fashion, animal, vlog, face and AR filter options. Animated stickers and music can also be posted to the user's profile. The platform's affordances are all geared towards mobile use and the seamless integration of smartphone and micro-video creation and consumption.

When you open the app, the default newsfeed is an algorithmic curation of video-based content called *For you*. Unlike Facebook or Instagram, for example, the user does not need to actively search or follow particular accounts before seeing content. Thus, while you might expect this home screen to be empty when a user has not engaged with the app before, the content stream is always ready, "instantly entertaining", as TikTok's marketing asserts. The feed works more like Netflix's homepage and adjusts in real-time to your habits and responses, including location, the length of time you watch any particular video, and your likes, searches, shares and posts. You navigate the feed, perfectly suited to mobile use, by scrolling vertically through an endless stream – an *infinite digital arcade* – of videos lasting 15 to 60 seconds, with a mix of pranks, gags, advertising, skits, lip-syncing and dancing. Down the right side of the screen, five icons are visible: a small circle with a profile picture of the video poster; a love heart, for "liking" content; a speech bubble, for comments; a share button; and a small turning record, which allows users to link to other videos using the same music. Users can comment on all public accounts, but can only comment on a private account if they are friends. You can also hold a finger down on a particular video to indicate that you wish to see less of its kind of content. The amount, timing and content of advertising varies between users, despite similar demographics and locations.

TikTok is an endless stream of short, entertaining memes and *human GIFS* (Poniewozik et al., 2019) set to music, perfectly suited to viewing via a smartphone. While we know as researchers that the short clips have been algorithmically curated by an AI constantly assessing our interactions, the experience is so seamless that you lose sight of this mediation. As noted in TikTok's marketing: "the more you watch, the better it gets".

The platform's content stream is rendered endless in a number of ways. There is no gap between videos, either visually or time-wise. Videos play automatically and are set to loop unless you tap to exit or pause. Further, the platform does not include time and date information on individual videos. This means that users are more likely to engage with older content (something we tend to avoid doing on other platforms, such as Facebook) and help it to go viral (Matsakis, 2019b). Newness and timeliness are less of an advantage. The only way to find out when a video was posted is to go to the creator's homepage. This timelessness means that time-dependent content such as news and events is less likely to become popular and is thus dis-incentivised for creators. This factor also means that content remains aligned with the platform's emphasis on an "apolitical" environment of "creativity and joy", rather than one that includes politics, conflict or controversy.

To populate the non-default *Following* feed, you can choose to follow users pushed to you via *For you*, or actively search for content on the *Discover* page, which presents trending hashtags along with a search tool. Many of these hashtags constitute *challenges*, a key organising force of TikTok's content ecology and user engagement, and another vital contributor to the sense of endlessness engineered into the app. These challenges are set by the platform or a user, or are sponsored by a particular brand. Indeed, responding to a challenge, recording a duet or a reaction video, or lip-syncing to someone else's audio are the dominant means through which users connect to each other, beyond the familiar liking or commenting. These particular affordances encourage users to connect via sharing, replicating, extending and collaborating on the content itself, in contrast to older social media platforms that emerge out of organic social networks and mediated forms of commentary. The user behaviour on TikTok also means that content can go viral without a user having many followers, which introduces a democratising effect to the app, no matter how illusionary this might be, given the aggressive algorithmic management engineered into the process.

On TikTok, then, content *is* connection. Content becomes the means through which users are encouraged to engage and the network is built, with algorithmic processes driving and organising its emergence as an endless stream of shared responsivity. Social conversation and connections are secondary to this process, displacing the sense of public engagement afforded by platforms that rely on a more traditional logic of "following" and "friends" to organise the experience. The app's algorithmic processes feel more opaque in their workings, less tethered to organic social networks and more natively and aggressively "promotional" than older platforms such as Facebook.

Content that lends itself to this logic – sharable, repeatable, context-independent, able to be added to or remixed – is dramatically favoured by such a media environment. As a further consequence of this emphasis, "TikTok all but eradicates traditional norms about cultural ownership" (Poniewozik et al., 2019, para. 23), giving value to copies, parody, remixing and referentiality over traditional notions of creative originality, source material or authorial intent. And this emphasis is engineered into the options users are given for video creation in a host of ways.

To create a video, users tap the plus sign at the bottom of the screen. Videos can be made in response to another user's content via *Share* if that other user has made options available to *duet* or *react*. Videos are a maximum of 15 seconds long, but there is an option to connect a few together for up to 60 seconds in total. Users can also upload longer videos if they were recorded outside the app.

Once the user is in camera mode, they can start recording immediately or first choose an audio track, the best option if the video is going to involve lip-syncing or dancing to a particular track. The *Sound* icon gives the user a menu reminiscent of Spotify, from which to browse and choose a song among a range of categories, including Trending, Country, Hip Hop and Featured. It is not possible to choose which section of a particular song you want; these segments are pre-determined and tend to be a catchy section that lends itself to lip-syncing, dancing or some of kind of video-able action or replicable micro-narrative. Not all songs will suit and this has an impact on the kind of content being produced and becoming successful if using TikTok as a promotional tool. Users can also upload their own audio content by using it on an uploaded video, which others can then access. This is how Lil Nas X's riff from "Old Town Road" went viral and that track is typical of the kind of music that does well: highly repetitive, quirky and simple. The song menu will also lead you to the tracks for particular TikTok challenges.

There are options to manipulate the audio-visual text during and after the video-capture process and most require some practice to use well. A user can adjust recording and playback speed. They can also add transitions, jump-cuts, green-screen backgrounds, text and a wide variety of AR filters for the face or environment. The effect of incorporating TikTok's video manipulation tools is often a far greater and highly entertaining theatricality and playfulness than audio-visual content produced on a platform such as YouTube or Instagram. A typical two-step micro-narrative structure has emerged, common to many TikTok challenges in particular: viewers have learned to anticipate a now familiar structure involving a "set-up" and then the punchline or transformation, with a jump-cut or transition in between, designed to keep the viewer engaged until the end (Bresnick, 2019). This, Bresnick argues, is central to the architecture and experience of the app as an *intensified virtual playground*, one that combines the "intensification of contemporary cinematic techniques with the changing characteristics of play on mobile devices" (p. 2).

Finally, when a user does want to extract themselves from this infinite digital arcade of endless micro-memes set to music, leaving is fairly straightforward. And, perhaps surprisingly, so is closing one's account. Indeed, involuntary loss of one's account has been an issue for thousands of users caught up in controversies around under-age accounts, privacy and ByteDance's heavy-handed public relations response. Unlike Facebook, famous for how difficult it makes leaving and the fact that it retains all of your data and images, TikTok claims to delete the videos produced by a user 30 days after the closure of the account.

DISCUSSION: TIKTOK AND THE FUTURE OF CREATIVE WORK

Our analysis of TikTok's socio-technical architecture was designed to examine the app's socio-economic and cultural context; its engineering of user experience; and how creativity and playfulness are transformed in the process. Alongside our discussion of the changing nature of both professional and amateur creative labour that introduced this chapter, the analysis highlights a number of issues directly related to questions of creative work and an increasingly algorithmic future.

TikTok exemplifies the many ways in which we now delegate the work of *reassembling the social* (Latour, 2005) – "the sorting, classifying and hierarchizing of people, places, objects and ideas" (Striphas, 2015, p. 395) – to carefully engineered and commercially orientated algorithms. As such, this cultural management is now increasingly organised by a logic of surveillance, data aggregation and monetisation. As a result, predictive correlation is displacing direct human judgement and understanding (Andrejevic, 2014), even in the domains of aesthetics and play. A minor but telling symptom of this alienation of participants from their own datafied actions is the current trend on TikTok to include the #foryou hashtag on all uploaded videos, on the suspicion that this will improve the chances of being promoted by the algorithm and going viral. This clearly illustrates the two-way nature of the relationship between algorithms and our culture – our habits of thought and cultural expression are being increasingly shaped by the logic of these algorithms (Striphas, 2014): "Culture now has two audiences, in other words: people and machines. Both will have a significant hand in shaping the material that finds its way into the public realm" (Striphas, 2014, "How do you envision the future of the Cultural Industry?", para. 4).

TikTok is designed with the circulation of a particular type of content in mind: playful, meme-based, visually engaging, reproducible and apolitical. The platform also works relentlessly to render public what might have once been regarded (and is still by many) as deeply cringeworthy, private and amateur activities: "TikTok is every embarrassing thing you do alone in your bedroom, but broadcast for the whole world to see" (Tait, 2019, para. 3). And, as ByteDance controls the visibility of users, it remains unclear which videos achieve wide visibility. In fact, TikTok was recently criticised in the media for a policy that limited the number of views of videos posted by disabled users, which TikTok claimed would leave them susceptible to bullying (Perper, 2019). In other cases, some TikTok users attempt to "game" the app's algorithm, developing a range of strategies to generate awareness about various causes. For example, an American teen recently posted a video to raise awareness about the treatment of Muslims in Chinese camps that was topped and tailed with a beauty tutorial (Kelion, 2019). While Douyin does moderate political content, TikTok claims it does not (Kelion, 2019) and the operations of its algorithms remain unavailable for public or regulatory scrutiny. What is at stake, here, is the privatisation of decisions determining the "values, practices, and artifacts – the culture, as it were – of specific social groups" (Striphas, 2015, p. 406). Questions of user privacy and surveillance are particularly difficult to address or regulate in such an environment. Nevertheless, the clever use of a make-up tutorial to embed and circulate a political message is an equally instructive instance of creativity *within* the affordances provided by the platform itself. So, too, Lil Nas X's savvy use of the #yeehaw challenge to promote his own music. And these two examples highlight the paradox at the heart of TikTok itself: the platform relies on a user freedom and creativity to drive engagement, teach the algorithms and build its network; yet, this very creativity must be subject to control in response to the apolitical vision and regulatory context in which it functions. The many controversies dogging ByteDance in recent months serve to demonstrate the difficult dance required to balance these imperatives and the two contradictory models of creativity on which they rely.

It has always been in an organisation's interest to convert as much ambiguous, relational *embedded* capital as possible into *separable* capital, thus reducing an organisation's reliance on individual and intangible skills, relationships and knowledge (Bowman and Swart, 2007). An app such as TikTok might be understood as the most recent form of a digital, post-industrial iteration of this logic: TikTok's architecture is designed for the efficient *datafication* of embodied, private and ephemeral experiences and – in this case – playful creativity and connection. Indeed, a recent study of video-game commentary on YouTube (Postigo, 2016) concludes similarly that "all forms of cultural practice traversing through architectures framed by algorithms and affordances are similarly captured and converted to inventory and enter the organizational logics of platform owners" (p. 335). TikTok and the kind of creative labour encouraged by its affordances and algorithmic curation are typical of online culture in so far as it is driven by the *circulation of affect*, spectacle and the experiential (Dean, 2013). More specifically, the app exemplifies Dean's (2016) notion of *secondary visuality*, which helps to explain the particular power of the hypervisual nature of the content under production and circulation on TikTok and the dynamic through which imagery circulates *as* connection outlined above:

> Because images circulate as conversations, we find ourselves engaging in a new communicative form where *the originality or uniqueness of an image is less important than its common, generic qualities, the qualities that empower it to circulate quickly and easily, that make it contagious.* Images function as visual colloquialisms, figures of speech, catchphrases and slang….What matters is whether an image is repeated, whether it incites imitation, whether it can jump from one context to another. An image's circulatory capacity, its power to repeat, multiply and acquire a kind of force, has triumphed over its meaning….*We live montage.* (p. 4, emphasis added)

These characteristics have clear implications for the future of professional forms of creative labour as much as they do for the nature of user-generated content under production. To succeed in such a system, creatives must not only understand the role algorithms play in content circulation and engagement, and craft their content accordingly; they must also be prepared to adapt to adjustments and fine-tuning of the algorithms in response to commercial and regulatory imperatives.

CONCLUSION

TikTok represents a shift in how the products of creative labour have been transformed. The game of pretending to be a pop star in the 1980s was beyond the reach of institutional management and commodification of play, even if we made use of commercial artefacts, such as music tracks on records or cassettes, to scaffold and organise our performances of femininity and fame. TikTok, on the other hand, operates as a *digital enclosure* (Andrejevic, 2007) designed to capture and transform creative play into a promotional apparatus and form of digital labour, with algorithms as the organising force. And as our discussion of the kinds of videos produced and shared on TikTok illustrates, such a system favours creative content with the capacity to generate engagement, go viral and produce user data; the particular cultural, aesthetic or social qualities of said content are radically de-emphasised. In the process, the value of this creative labour is thus disassociated from particular bodies and communities, and dispersed more broadly throughout a culture and economy profoundly mediatised by *communicative capitalism* (Dean, 2005). To work within such a system, as Lil Nas X illustrates, is about developing platform literacy and a willingness to play the algorithms at their own game and leverage their logic of virality towards other, more potentially lucrative and independent creative practices.

In an age of mass digital communication, re-enacting a music video generates value for digital platforms that benefit from bringing together private experiences with a media infrastructure. These activities can be properly understood as a kind of unpaid digital labour, in so far as they produce value for the platforms on which they occur. In contrast to the experience of lip-syncing to pop stars back in the 1980s, TikTok provides a suite of tools designed to enhance the creativity of the user. It also comes with a model that promotes and broadcasts amateur performances to a global audience. Another change is that in the 1980s we could re-enact and re-stage our performances for hours on end with no "final product", but TikTok shapes creative play into a technically sophisticated media document of between 15 and 60 seconds. Users experience forms of connection through their creative outputs that are largely mash-ups and remixes of cultural texts. There is a merging of everyday communication practices within a larger flow of images, experiences and emotions that are used in combination with stamps, music and AR effects. Dean (2016) argues that this flow of secondary visuality should be understood as a form of collectivity, and we support this perspective. The duets shared and liked by multiple users on TikTok and the endless popularity of cat videos are a "collective expression of common feelings" that extend beyond the individual to the many (Dean, 2016, p. 5).

Rather than dismissing the circulation of TikTok content as an inconsequential form of expression, it should be acknowledged as part of a larger segment of communication that involves the compression of media with creative production. This is significant because the circulation of creative output becomes a critical element of the work-making process in which amateur forms of creativity are re-purposed into a form of digital labour. This differs from previous forms of creativity and play. Like other generations, young people are resourceful in their use of the platform, crafting from its suite of resources sophisticated media documents. Their canny and creative work within the *integrated circuit* (Haraway, 1987) of TikTok's socio-technical architecture can provide an important map for the work of other creatives.

NOTES

1.   ORCID iD: 0000-0002-7825-9018, School of Communication and Arts, University of Queensland, n.collie@uq.edu.au.
2.   ORCID iD: 0000-0003-1062-5525.

3.   See https://www.youtube.com/watch?v=N7yu-KMJ_iI&t=84s, for example.
4.   A potentially lucrative practice of manufacturing virality by selling mass retweets (often of stolen tweets) using Twitter's TweetDeck interface (Reinstein, 2018).
5.   The racial politics of this categorisation later played out, when Billboard dropped the single from its Hot Country Songs category after it went mainstream (Leight, 2019).
6.   Despite its intense popularity, however, Musical.ly is remarkably under-researched (Literat and Kligler-Vilenchik, 2019).

REFERENCES

Andrejevic, M. (2007). Surveillance in the digital enclosure. *The Communication Review*, **10**(4), 295–317.

Andrejevic, M. (2014). The big data divide. *International Journal of Communication*, **8**, 1673–89.

Arch (2019). How do TikTok gifts work? *TechJunkie*, 24 May. Retrieved from https://www.techjunkie.com/how-do-tiktok-gifts-work/.

BBC (2019). TikTok owner scrutinised over Musical.ly deal. *BBC*, 1 November. Retrieved from https://www.bbc.com/news/technology-50267985.

BeautyTech.jp (2018). How cutting-edge AI is making China's TikTok the talk of town. *Medium*, 18 November. Retrieved from https://medium.com/beautytech-jp/how-cutting-edge-ai-is-making-chinas-tiktok-the-talk-of-town-4dd7b250a1a4.

Beer, C. (2019). Is TikTok setting the scene for music on social media? 3 January. Retrieved from https://blog.globalwebindex.com/trends/tiktok-music-social-media/

Bowman, C. and Swart, J. (2007). Whose human capital? The challenge of value capture when capital is embedded. *Journal of Management Studies*, **44**(4), 488–505.

Bresnick, E. (2019). Intensified play: Cinematic study of TikTok mobile app. Retrieved from https://www.academia.edu/40213511/Intensified_Play_Cinematic_study_of_TikTok_mobile_app.

ByteDance (n.d.). Research areas at ByteDance. Retrieved from https://bytedance.com/en/technology.

Chow, A.R. (2019). Lil Nas X on "Old Town Road" and the Billboard controversy. *Time*, 30 March. Retrieved from https://time.com/5561466/lil-nas-x-old-town-road-billboard/.

Dean, J. (2005). Communicative capitalism: Circulation and the foreclosure of politics. *Cultural Politics*, **1**(1), 51–74.

Dean, J. (2013). *Blog theory: Feedback and capture in the circuits of drive*. Hoboken, NJ: John Wiley & Sons.

Dean, J. (2016). Faces as commons: The secondary visuality of communicative capitalism. *Open! Platform for Art, Culture, and the Public Domain*. Retrieved from https://onlineopen.org/faces-as-commons.

Hamilton, I.A. (2019). TikTok was bigger than Instagram last year after passing the 1 billion download mark. *Business Insider Australia*, 28 February. Retrieved from https://www.businessinsider.com.au/tiktok-hit-1-billion-downloads-surpassing-instagram-in-2018-2019-2?r=US&IR=T.

Haraway, D. (1987). A manifesto for cyborgs: Science, technology, and socialist feminism in the 1980s. *Australian Feminist Studies*, **2**(4), 1–42.

Harwell, D. and Romm, T. (2019). TikTok's Beijing roots fuel censorship suspicion as it builds a huge U.S. audience. *The Washington Post*, 16 September. Retrieved from https://www.washingtonpost.com/technology/2019/09/15/tiktoks-beijing-roots-fuel -censorship-suspicion-it-builds-huge-us-audience/.

Hern, A. (2019a). "Adults don't get it": Why TikTok is facing greater scrutiny. *The Guardian*, 5 July. Retrieved from https://www.theguardian.com/technology/2019/ jul/05/why-tiktok-is-facing-greater-scrutiny-video-sharing-app-child-safety.

Hern, A. (2019b). Revealed: How TikTok censors videos that do not please Beijing. *The Guardian*, 25 September. Retrieved from https://www.theguardian.com/technology/ 2019/sep/25/revealed-how-tiktok-censors-videos-that-do-not-please-beijing.

Kelion, L. (2019). Teen's TikTok video about China's Muslim camps goes viral. *BBC News*, 26 November. Retrieved from https://www.bbc.com/news/technology-50559656.

Latour, B. (2005). *Reassembling the social: An introduction to Actor-Network Theory*. New York, NY: Oxford University Press.

Leight, E. (2019). Lil Nas X's "Old Town Road" was a country hit. Then country changed its mind. *RollingStone*, 26 March. Retrieved from https://www.rollingstone .com/music/music-features/lil-nas-x-old-town-road-810844/.

Light, B., Burgess, J. and Duguay, S. (2018). The walkthrough method: An approach to the study of apps. *New Media & Society*, **20**(3), 881–900.

Literat, I. and Kligler-Vilenchik, N. (2019). Youth collective political expression on social media: The role of affordances and memetic dimensions for voicing political views. *New Media and Society*, **21**(9), 1988–2009.

Lorenz, T. (2019). TikTok stars are preparing to take over the internet. *The Atlantic*, 12 July. Retrieved from https://www.theatlantic.com/technology/archive/2019/07/ tiktok-stars-are-preparing-take-over-internet/593878/.

Matsakis, L. (2019a). FTC hits TikTok with record $5.7 million fine over children's privacy. *Wired*, 27 February. Retrieved from https://www.wired.com/story/tiktok-ftc -record-fine-childrens-privacy/.

Matsakis, L. (2019b). On TikTok, there is no time. *Wired*, 3 March. Retrieved from https://www.wired.com/story/tiktok-time/.

Nicas, J., Isaac, M. and Swanson, A. (2019). TikTok said to be under national security review. *New York Times*, 1 November. Retrieved from https://www.nytimes.com/ 2019/11/01/technology/tiktok-national-security-review.html.

Niewenhuis, L. (2019). The difference between TikTok and Douyin. *SupChina*, 25 September. Retrieved from https://supchina.com/2019/09/25/the-difference -between-tiktok-and-douyin/.

Parkin, B. (2019). Tiktok: India bans video sharing app. *The Guardian*, 17 April. Retrieved from https://www.theguardian.com/world/2019/apr/17/tiktok-india-bans -video-sharing-app.

Perez, S. (2019). TikTok explains its ban on political advertising. *Tech Crunch*, 4 October. Retrieved from https://techcrunch.com/2019/10/03/tiktok-explains-its-ban -on-political-advertising/.

Perper, R. (2019). TikTok admits it hid disabled users' videos appearing on the app's main feed, claiming the "blunt" policy was used to prevent bullying. Business Insider, 3 December. Retrieved from https://www.businessinsider.com.au/tiktok-limited -disabled-users-videos-from-main-feed-anti-bullying-2019-12?r=US&IR=T.

Poniewozik, J., Hess, A., Caramanica, J., Kourlas, G. and Morris, W. (2019). 48 hours in the strange and beautiful world of TikTok. *New York Times*, 10 October. Retrieved from https://www.nytimes.com/interactive/2019/10/10/arts/TIK-TOK.html.

Postigo, H. (2016). The social-technical architecture of digital labor: Converting play into YouTube money. *New Media & Society*, **18**(2), 332–49.

Reily, D. (2019). Lil Nas X's "Old Town Road" was destined to disrupt. *Vulture*, 4 April. Retrieved from https://www.vulture.com/2019/04/lil-nas-x-old-town-road -controversy-explained.html.

Reinstein, J. (2018). "Tweetdecking" is taking over Twitter. Here's everything you need to know. *Buzzfeed News*, 12 January. Retrieved from https://www .buzzfeednews.com/article/juliareinstein/exclusive-networks-of-teens-are-making -thousands-of-dollars.

Rettberg, J.W. (2017). Hand signs for lip-syncing: The emergence of gestural language on musical.ly as a video-based equivalent to emoji. *Social Media + Society*. doi:10 .1177/2056305117735751.

Savic, M. and Albury, K. (2019). Most adults have never heard of TikTok. That's by design. *The Conversation*, 11 July. Retrieved from https://theconversation.com/most -adults-have-never-heard-of-tiktok-thats-by-design-119815.

Simsek, B., Abidin, C. and Brown, M.L. (2018). musicl.ly and microcelebrity among girls. In C. Abidin and M.L. Brown (eds), *Microcelebrity around the globe: Approaches to cultures of internet fame* (pp. 47–56). Bingley: Emerald Publishing.

Singth, M. (2019). TikTok tests social commerce. *TechCrunch*, 15 November. Retrieved from https://techcrunch.com/2019/11/15/tiktok-link-bio-social-commerce/.

Strapagiel, L. (2019). How TikTok made "Old Town Road" become both a meme and a banger. *Buzzfeed News*, 8 April. Retrieved from https://www.buzzfeednews.com/ article/laurenstrapagiel/tiktok-lil-nas-x-old-town-road.

Striphas, T. (2014). Algorithmic culture. "Culture now has two audiences: people and machines": A conversation with Ted Striphas. (G. Granieri, interviewer). *Medium*, 1 May. Retrieved from https://medium.com/futurists-views/algorithmic-culture -culture-now-has-two-audiences-people-and-machines-2bdaa404f643.

Striphas, T. (2015). Algorithmic culture. *European Journal of Cultural Studies*, **18**(4–5), 395–412.

Tait, A. (2019). TikTok has created a whole new class of influencer. *Vice*, 15 May. Retrieved from https://www.vice.com/en_in/article/597dbn/tiktok-what-uk-stars-influencer.

Terranova, T. (2000). Free labour: Producing culture for the digital economy. *Social Text*, **18**(2), 33–58.

TikTok (n.d.). Our mission. Retrieved from https://www.tiktok.com/about?lang=cs.

Tolentino, J. (2019). How TikTok holds our attention. *The New Yorker*, 23 September. Retrieved from https://www.newyorker.com/magazine/2019/09/30/how-tiktok-holds -our-attention.

van Dijck, J. (2013). *The culture of connectivity: A critical history of social media*. Oxford, UK and New York, NY, USA: Oxford University Press.

Verberg, G. (2019). Are Douyin and TikTok the same? *What's on Weibo*, 9 January. Retrieved from https://www.whatsonweibo.com/are-douyin-and-tiktok-the-same/.

Wodinsky, S. (2019). TikTok confirms it's testing shoppable videos. *Adweek*, 14 November. Retrieved from https://www.adweek.com/digital/tiktok-confirms-its -testing-social-shoppable-videos/.

12. Managing embedded creative work: the challenge of causal ambiguity

Cliff Bowman[1] and Juani Swart[2]

INTRODUCTION

We consider the role of causal ambiguity in the management of those involved in creative work. *Causal ambiguity* refers to a lack of understanding and insight into how things work, specifically about the relations between cause and effect, or actions and outcomes. It affects a manager's ability to comprehend fully both the nature of skilled and creative activity and the links between an activity and firm performance. Here, we unpack ambiguity, explore how it creates challenges for the management of creative talent and teams, and suggest practices that could reduce some negative impacts of causal ambiguity. We focus on creative workers who are embedded in organisations that do not have creativity as their core activity; that is, those outside the creative industries.

We begin with a brief exploration of the nature and qualities of creative work.

CREATIVE WORK

Creative work would clearly include the visual and performing arts, cinema and television films, sound recordings, book publishing and games development, and these activities have some common attributes, which Caves (2000) describes as follows:

1. Demand for the outputs of creative work is often uncertain.
2. Creative workers tend to care about the "product" they create.
3. Whilst some outputs can be produced by a single person (e.g., a painting or a novel), many require diverse skilled and specialised workers interacting in loose teams that are formed to produce a specific one-off product (e.g., a film or video game) and these teams disband when the project is completed.

4. Creative products are "one-offs" that are evaluated by consumers against other similar but differentiated products.
5. Performers are often "ranked" (e.g., "A" list actors) and their rankings and reputations will fluctuate.
6. Time is of the essence in many creative processes: the film has to be shot in these weeks or the concert takes place on this date. Coordination of those creatives and other complementary and essential supporting activities is a critical quality of most creative endeavours.
7. Some creative products are one-offs (e.g., the concert), but others can deliver income streams over many years through royalty payments.

To these artistic pursuits, we can add other creative activities that display many of these seven qualities, such as advertising, fashion, interior design and architecture. But creative work stretches beyond artistic endeavours, for example, industrial design, large-scale engineering projects, management consulting interventions, in-house advertising and communication projects, and all forms of research. This type of work corresponds with the categories of intangible investments made by firms, and the level of this investment has exceeded the level of tangible investments in the United Kingdom (UK) since the early 2000s (Haskell and Westlake, 2018). These creative activities typically produce unique, "one-off" outcomes and they often involve teams of highly skilled professionals who interact to produce unique solutions to complex problems.

It is this wider definition of creative activity that we focus on in this chapter, exploring the managerial challenges this work presents. Whereas the obviously artistic pursuits listed above have evolved appropriate managerial practices that recognise the peculiarities of artistic work, our extended list of creative activities typically occurs within contexts where creative work may not be the focus of the organisation. Nevertheless, the need to "manage" this creative work appropriately can often be critical to the success of the wider organisation. The reason for this is that, within these organisations, creative work is often responsible for producing differentiated outputs that are a major source of competitive advantage for the firm. Those performing these activities may themselves be unique and valuable resources for the firm, or they help create these resources.

We distinguish between those engaged in the creative industries listed earlier and those engaging in creative work inside firms in non-creative industries. Hearn and Bridgstock (2014) refer to these employees as *embedded creatives*. Bowman and Swart (2007) distinguish between three forms of capital: *separable* capital (equipment, patents, brand names); *embodied* capital (skills, knowledge embodied in individuals); and *embedded* capital, which are the synergistic relationships and interactions between separable and embodied forms of capital. An understanding of all three of these forms of capital is relevant to the work of embedded creatives, but embedded capital may be the most crucial for the future work of embedded creatives.

We are interested in actions that impact the management of these embedded creatives. Causal ambiguity influences decision-making and action. Ambiguity can lead to sub-optimal decisions and choices. High levels of ambiguity can lead to errors in the management of creative human resources that might be the outcome of either inappropriate interventions or inaction. These errors can lead, inadvertently, to the destruction of creative resources. Thus, we next set out a brief explanation of the resource-based view of the firm, which acts as a theoretical base for the chapter.

THE RESOURCE-BASED VIEW OF THE FIRM

The resource-based view of the firm (RBV) explains how inimitable resources are the source of superior firm performance. These resources can be different types of "human capital", or they can be brands, patents, equipment, reputation, locations and so on. To qualify as a resource, they have to be simultaneously valuable, rare, inimitable and non-substitutable, the "VRIN" criteria (Barney, 1991; Hall, 1992). Firms gain and sustain competitive advantage through the possession of valuable and inimitable resources that generate *rents*, initially captured by the firm, but which may be passed on to those who supply the scarce resources (Coff, 1999).

The value of human capital is in the knowledge and skills that individuals accumulate. Some knowledge, for example, an assembly routine in the manufacture of phones, can be codified; that is, the sequence of tasks involved in the assembly of the phone can be described and explained. In contrast, *knowhow*, embodied in individual skills and collective tacit routines, resists codification. *Creative work involves individual and collective knowhow, which will be resistant to codification, context specific, socially complex and causally ambiguous* (Boon et al., 2018; Dierickx and Cool, 1989; Lippman and Rumelt, 1982).

Early contributions to the RBV recognised the role of complexity as a barrier to rival firms imitating valuable resources, specifically human resources. Complex processes, practices and relationships make it difficult for rival firms to comprehend and replicate valuable knowhow embedded in individuals and teams (Bowman and Swart, 2007). The interactive and processual nature of value creation typically displays all the qualities of *complex tasks*: the presence of (a) multiple pathways to arrive at a desirable outcome (equifinality); (b) many desirable outcomes; (c) conflicting inter-dependencies in these pathways to outcomes; and (d) uncertainty in pathway to outcome linkages (Campbell, 1988). Resource complexity may prevent rival firms from imitating valuable human capital, but it also poses challenges to managers *inside* the firm. Managers may well experience "internal" causal ambiguity, which would restrict their ability to protect, deploy or recreate complex creative resources. Moreover, resource complexity also suggests that resource-based advantages may well not result from *deliberate* strategising; they are more likely to emerge from "unmanaged" processes (Mintzberg and Waters, 1985).

There has been a tendency in the RBV to treat resources as discrete components in the firm's value system that can be identified, valued and potentially traded (Kraaijenbrink et al., 2010). We need to extend this "component" view of resources in order to fully comprehend the nature of creative resource-based advantages. *We need an approach that sees competitive advantage as deriving from talented individuals and as inherent in emergent webs of interactions between creative people and those with complementary and enabling capabilities.* Thus, creative work is typically *complex*, and this generates causal ambiguity, which we now explore.

CAUSAL AMBIGUITY

An individual's experience shapes how they perceive the context of a firm. According to Hambrick and Mason's upper echelons perspective (Hambrick, 2007; Hambrick and Mason, 1984), managers engage in selective perceptions of the wider environment, and their interpretations are influenced by their cognitions and values. These interpretations inform the actions managers undertake. Managers engage in sense-making, which serves as a springboard to action (Gioia and Chittipeddi, 1991; Weick et al., 2005).

Whilst there are many facets to the social perception of reality, here we focus on *ambiguity*. *Causal ambiguity* refers to a lack of understanding and insight into how things work, specifically about the relations between cause and effect, or actions and outcomes (King, 2007; King and Zeithaml, 2001; Reed and De Fillippi, 1990). Ambiguity is subjectively experienced. For instance, someone new to the organisation may have little understanding of the value creation processes enacted by their new colleagues, whereas among those with many years' experience, there is likely to be a shared and accurate understanding of how the system works. Causal ambiguity is a function of interactions between a context (e.g., a firm) and an observer, for example a manager. Causal ambiguity is subjectively experienced by managers and will affect managers' actions. These actions in turn are influenced by the context and also influence the context. We propose that this dynamic has important implications for management of embedded creatives, particularly human resources (HR) management of creative talent and teams.

King and Zeithaml (2001) distinguish between two forms of causal ambiguity: characteristic and linkage ambiguity.

Characteristic ambiguity refers to circumstances where an observer of an activity is unable to fully understand how the activity is performed. This could be due to the tacit knowhow involved in the performance, which resists direct comprehension and codification. The performers themselves may also be unable to articulate or explain how they do what they do. Social complexity compounds characteristic ambiguity, particularly where the value creation process involves multiple ongoing interactions between flexible teams of people who all bring their own particular knowhow to the collective endeavour.

Linkage ambiguity occurs where an observer is unable to clearly identify the direct causal links between an activity and firm performance. Complexity in the firm's value creation processes would exacerbate linkage ambiguity, as would an observer's lack of experience of the firm or the firm's industry. Ambiguity is an outcome of the *interaction* between the manager and the context.

King (2007) has drawn together the primary *drivers* of characteristic and linkage ambiguity (see Figure 12.1):

- *Tacitness*: Individual and collective competences display tacitness where it is difficult for those involved to articulate, describe or explain to others how they do what they do. Creative skills resist codification, and hence they generate characteristic ambiguity from the perspective of someone seeking to manage this activity.
- *Distance* refers to the spatial gap between the observer and the activities involved in a creative process. The bigger the gap, the less likely the observer is to fully comprehend either the activity or its impact on firm performance. Hence, distance impacts both characteristic and linkage ambiguity.
- *Complexity* refers to the number of variables that are involved in a competence and how they are related. The more variables involved, the more complex the competence, and complexity increases further where these variables interact in non-linear ways. Complexity impacts both forms of ambiguity.
- *Time lag*: The longer the time period between a creative activity and the performance outcomes of this activity, the greater the degree of linkage ambiguity experienced by the observer. Decision-makers' understanding of causal relationships distorts and deteriorates over time (Walsh and Ungson, 1991).

Figure 12.1 provides a holistic representation of causal ambiguity and its drivers, drawing together the concepts of King and Zeithaml (2001) and King (2007). It can be seen that *complexity* contributes to both forms of ambiguity, whereas *tacitness* primarily contributes to characteristic ambiguity. For

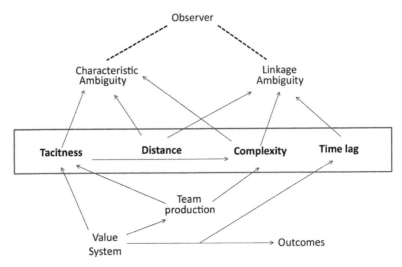

Source: Authors drawing on concepts in King (2007) and King and Zeithaml (2001).

Figure 12.1   The drivers of characteristic and linkage ambiguity

example, the creation of a television advertisement is typically a complex process involving multiple actions and interactions among a group of individuals with differing but complementary skills. Some would be full-time employees in the corporation's marketing department, some would be working in an advertising agency, and others would be freelancers involved in shooting the advertisement and post-production work. *Team production* in creative activities is the norm, and the nature and quality of interactions resist attempts to codify them. Those involved in team production are drawing on their knowhow, which has been built through their experience. Moreover, how they interact one with another will be an example of a collective *tacit routine,* in so far as the nature, timing and effects of these multiple interactions (e.g., in an initial meeting) will not have been explicitly designed and will emerge in the moment.

Figure 12.1 indicates that the *distance* of the observer from the activity increases both forms of ambiguity. Distance refers not just to spatial separation, but also to hierarchical distance experienced, for example, by an executive who has little interaction with "shop floor" employees. Moreover, cognitive distance will be high where the activity being observed is outside the manager's past experience. The manager's worldview, shaped by their experience, means that they may struggle to fully comprehend the activities and their wider context.

The *time lag* between the activity in the system and the ultimate performance outcome affects linkage ambiguity. Attenuated value creation processes compound the problem of identifying the causes of firm performance.

Managers' perceptions and interpretations of the social world they inhabit will be shaped by their experience. Where a group of managers spend their career within one industry, they will be likely to form similar worldviews, and these often unspoken assumptions about the world will guide their discourse and behaviours. These shared worldviews will reflect the nature of the industry and occupations or functions the managers work in. We might expect a group of managers in the auto industry to have similar worldviews, and these might be different to managers from the food industry, or those in textiles, or construction.

The aestheticisation of goods and services (Harvey, 1989; Jameson, 1991) has led to an increasing reliance on creative work to improve the functional *appeal* of products. Whilst the core product may remain unchanged, the product *surround* is continually developed through creative work. This "soft innovation" (Stoneman, 2010) is increasingly performed by embedded creatives, who may interact with people working in other "full-time" creative firms, for example, advertising agencies, television production companies or graphic designers. But whilst those in full-time creative firms will experience organisation cultures and managerial worldviews that have emerged in these special contexts, the embedded creatives will likely be managed through the dominant culture of the "host" firm, for example, auto making or pharmaceuticals.

Over time, competing firms have developed and imitated efficient production processes and supply chains. For example, we expect cars to be reliable and not to rust. A consequence is that sources of product differentiation and competitive advantage have shifted from the efficient production of, for example, a reliable car, towards the aesthetic qualities that the particular car brand evokes. Thus, the efficient production of a reliable car becomes an *order qualifying* criterion in the auto industry, not an *order winning* criterion. The "aesthetic performance" of the car becomes increasingly critical to firm success. The clear challenge for incumbent managers is that their dominant worldview and the organisation's culture are potentially increasingly out of sync with the firm's emerging sources of competitive advantage.

If we juxtapose high and low levels of characteristic ambiguity and linkage ambiguity (see Ambrosini and Bowman, 2010) we can conceptualise four different contexts which managers of embedded creatives must deal with, namely:

Context A: Low linkage ambiguity and low characteristic ambiguity. In these contexts there is low system complexity, which typically generates low levels of characteristic and linkage ambiguity. Low complexity firm contexts can be readily understood, and managerial interventions can be selected and introduced with predictable outcomes.

Context B: Low linkage ambiguity and high characteristic ambiguity. In these contexts there are activities that involve significant levels of tacit knowledge that generate characteristic ambiguity, which increases the complexity and challenge of managing these processes.

Context C: High linkage ambiguity and low characteristic ambiguity. Here the tasks are readily codifiable but there are problems in identifying the links between any particular activity and overall firm performance; again, the managerial task becomes more challenging.

Context D: High linkage ambiguity and high characteristic ambiguity. Here high levels of complexity and tacit knowledge are present, which generate high levels of both linkage and characteristic ambiguity. We hypothesise that embedded creative work primarily occurs in these contexts.

We now explore some implications of these four contexts.

Context A: Low Characteristic Ambiguity (CA), Low Linkage Ambiguity (LA)

In these contexts we would locate activities that the observer can readily comprehend. The observer can also see the connections between the activity and the overall performance outcomes for the firm. The activity would be likely to consist of codifiable routines and the low system complexity would mean it is easy for the observer to see the performance impact of the activity. Low-skill, routinised work where employees interact with capital equipment in the process of value creation would be examples of these contexts, such as a delivery driver, a production-line operative, a shelf stacker in a supermarket, or a picker and packer in a warehouse. These activities are important, and if not performed appropriately there would be readily identifiable impacts on firm performance.

In these circumstances, managers perceive little ambiguity as the system is knowable. Managers can act with confidence that their understandings of the context are appropriate, and the relative simplicity of the system indicates that actions to improve performance would be relatively easy to identify, even if implementing these might be challenging.

Appropriate interventions might include:

- work simplification and de-skilling;
- automating wherever feasible;
- identifying, codifying and transferring "best practice";
- refining processes and procedures.

We would expect rival firms to have developed similar practices; thus, these would be "entry" requirements for this industry. Aligned managerial actions would refine and improve these systems, which should enable the firm to maintain a position in its market place.

Context B: High CA, Low LA

Here, the activity would be complex, involving tacit knowledge, which results in the observer experiencing high levels of characteristic ambiguity. That is, the observer might be able to recognise and appreciate the *outputs* of the activity, but they would not be able to understand how these outputs have been created. However, the overall value system is fairly easy to comprehend (low linkage ambiguity), enabling the observer to clearly identify the value contribution of the activity. For example, the firm may have outstanding individuals whose talents are recognised and demanded by the firm's clients. Managers may be unable to comprehend how these talented individuals do what they do (high characteristic ambiguity), but they are fully cognisant of their contribution to firm performance.

Complex creative tasks should encourage some *humility*, where managers acknowledge the limits of their comprehension and, if they decide to change the system, they should proceed with caution. Efforts to improve *comprehension* of the system's complexity will be worthwhile if the system is likely to persist in its current state.

Congruent management interventions for high characteristic ambiguity could include:

- the development of a culture that values knowledge creation and retention;
- trust-based relationships that value professional expertise;
- specialisation of tasks to facilitate the building of deep knowledge;
- decentralised structures that encourage decision-making by those with the greatest knowledge and who are closer to the problems and issues;
- reward systems that recognise and encourage the building of expertise.

Context C: Low CA, High LA

Here, the activity involves largely codifiable knowledge, but the connections between this work and firm performance are unclear. In a large complex value system involving thousands of employees across many functional areas, where within each area there are high levels of task specialisation, linkage ambiguity is likely to be high. The risks here result from inappropriate homogenising assumptions about all lower skilled codifiable activities. Consider this example. The firm manufactures and launders work-wear that is supplied to manufacturing firms. A decision was taken to outsource the delivery of laundered garments, the assumption being that this is a generic low-skill activity that an external delivery firm could do. As a consequence, the firm lost many clients. What had not been understood was that the delivery drivers did not just dump the laundered clothing and pick up the soiled garments; they each knew which clothing went to different departments and would distribute the garments across, in some cases, sites covering many acres. The drivers also understood which departments had higher risks of contaminated clothing, and dealt with these garments accordingly. None of this knowledge was available to the senior managers who were distanced from these activities and hence this knowledge could not be transferred to the outsourced delivery firm.

Appropriate management interventions for these contexts would be ones that try to improve the understanding of the system as a whole, and thereby try to reduce linkage ambiguity and use these insights to decide what to change. Practices that could help here would be cross-functional forums and building formal and informal relationships across the organisation. Clarity about the firm's strategy should help build an understanding of how disparate activities contribute collectively to deliver performance.

Context D: High CA, High LA

In these contexts, we have complexity and tacitness generating high levels of ambiguity; this is where most embedded creative work would be likely to be located. Managers may lack insights into the nature of skilled performance (high characteristic ambiguity) and they may also have limited understanding of how any particular employee or group's actions impact firm performance (high linkage ambiguity). Individual managers in these situations should *acknowledge complexity* and act accordingly. They should recognise the essential unknowability of the role of this work in the wider value system and engage with it in appropriate ways.

Congruent actions for this high ambiguity context could include facilitating:

- Cultural changes: "A culture which values importing knowledge from 'outside', and which encourages and rewards the generation of new ideas and promotes constructive collegial rivalry" and "management attitudes which encourage experiments, expects some 'failures' and recognises learning from them" (Bowman and Raspin, 2019, para. 11).
- Structural changes: "Forming of 'loose', fluid, temporary project teams which can generate new knowledge by enabling specialists with complementary knowledge to interact" and "joint ventures, alliances and cross-organisation collaborations" (para. 12).

- Systems changes: "Place many 'bets' but have timely appraisals so that failing projects can be 'killed' and resources can be readily redirected to more promising projects" and "frequent trialling of prototypes with high-quality feedback" (para. 13).

Complex contexts, which necessarily generate high levels of ambiguity, provide the most challenging circumstances for strategising, rendering analytical approaches virtually redundant. Strategy *emerges* from a series of trials, projects and experiments as the firm seeks to understand, shape and respond to the unfolding context.

In these contexts, the complex reality is matched by high levels of ambiguity. The complexity of these creative processes is recognised, and responded to appropriately by those responsible for managing the firm.

But problems arise when the reality indicates that high levels of ambiguity would be appropriate, but managers behave in ways that suggest they perceive a quite different version of reality. They perceive *low levels of ambiguity*, and, for example, treat creative work as if it was not socially complex, involving tacit knowledge. They engage *simplistically* with these activities and intervene inappropriately. The dangers here stem from inappropriate assumptions leading to the imposition of practices that turn out to be counterproductive.

To recap, we would expect in general that creative work would generate more ambiguity than other forms of productive activity. The primary reasons for this assumption would be the social complexity of many creative processes, and the tacit knowhow of creative people, which resists codification. Thus, creative work contexts generate ambiguity, which creates challenges for those charged with managing creative human resources.

As we explained, ambiguity is a function of the *relationship* between an observer and the context. Thus, ambiguity can change if either the observer or the context changes. For example, an HR manager who has spent her career in one creative firm, who was herself a "creative" and now finds herself in a senior role, would be likely to experience relatively low levels of linkage ambiguity. However, she may be less able to comprehend the nature of complementary creative knowhow contributions to a team effort; hence, she may still experience some degree of characteristic ambiguity in relation to her colleagues' work. But her familiarity with the creative process would probably lead her to acknowledge others' expertise and to act accordingly.

Contrast this with an HR professional employed in a non-creative context where there are a relatively small number of embedded creatives (e.g., the auto maker). His past experience equips him with knowledge of HR practices appropriate to car manufacturing contexts and which may have been successful. He may experience *low levels of ambiguity* if he believes these HR practices have general applicability – that is, they apply equally to creative work. These beliefs would cause him to engage confidently with these creative roles and to seek to apply these generic prescriptions. Thus he may act as if he were in context A, being unaware of the subtleties and complexities involved in embedded creative work.

The risks here are clear. An HR practice developed in one context (auto making) applied to another context (embedded creative work) will produce outcomes that cannot be predicted. Whilst the practice is "generic" (e.g., performance management), its specific deployment will interact with the particular nature of creative work roles, producing a *unique* outcome, not the desired or anticipated *generic* outcome. We explore some of these problems in the next section.

MISALIGNED ACTIONS

The dominant discourse surrounding the role of senior managers presumes they control "their" organisations, and they can predict the future sufficiently well to be able to predict the ultimate effects of any action they take. This narrative can result in complex value-creating systems being conceived of, and hence, managed, in *simplistic* ways.

Misalignments between the creative work context and managers' percep-tions provoke inappropriate managerial actions. These errors occur where psychological preferences for predictability and low ambiguity obscure and prevent insights into the complex reality. Whilst some thrive on uncertainty and ambiguity in some aspects of their lives, very few prefer this to be their dominant life experience. Preferences for predictability may cause managers to *enact* simplified realities (Smircich and Stubbart, 1985).

An example of inappropriate and simplistic responses to complexity would be *performance management*, where quantitative and easily measurable targets are set. But where the value creation process is complex, as it is in most embedded creative work, it is unlikely that simplistic targets will encourage the behaviours necessary to deliver valuable outcomes. Action, provoked by inappropriate key performance indicators (KPIs), will distort behaviour in unpredictable ways. These KPIs do not simplify, they merely mask the true complexity inherent in the system. Moreover, whilst the objective context requires engaged leaders, the simplification and denial of reality replaces leadership with *remote control*, delivered through target setting and moni-toring of outcomes, followed up by rewards or punishments. The impacts of these misaligned actions can have a devastating impact on subtle and complex creative processes.

Managers who lack insights into the creative process, or who lack confi-dence in their ability to exercise strategic leadership, may choose to implement *generic prescriptions* culled from popular management books. Similarly, executives may rely on external management consultants to guide their strategic actions. However, the consultants may be unable to discern the subtle and complex features of *this* particular context, or they may have a preferred "solution" that they have sold elsewhere. Their advice, if followed, could inadvertently destroy complex sources of advantage. Thus, a tendency to assume a *one size fits all* approach to management prescriptions may lead to improvements where the practice fits the context, but severe damage where the context and practice are misaligned.

Managers may join the firm from elsewhere and bring with them their "recipe". They may have developed and implemented this recipe in their previous organisation and judge it to have been successful. They may deem it appropriate for their new context, which may be quite different from their previous organisation. The required changes driven by this hubristic behaviour could be potentially damaging if the changes destroy subtle creative processes.

To sum up, generic practices, prescribed consultants' solutions or imported "recipes" from a manager's previous experience interact with the existing unique organisational context to produce an outcome that cannot be predicted.

There are numerous instances of large corporations acquiring small entrepreneurial start-ups in pharmaceuticals, management consulting and software development. The acquired business is required to adopt the practices and processes of the much larger and more bureaucratic acquiring corporation. The effect of the imposition of these practices is to drive out talented employees from the acquired firm. Having been used to a culture of decentralised and rapid decision-making, involvement, autonomy, flexibility and high levels of trust, they are unwilling to comply with the "top down" control culture of the acquiring corporation.

Whilst it is not possible to comprehend fully any complex social system, an awareness of the inherent complexity in the system is an important step in reducing the risks of inappropriate interventions. In addition, approaching complexity with some *humility* and an open mind can provide some important but partial insights into the nature of the value creation process. We explore these in the next section.

CAN CAUSAL AMBIGUITY IN EMBEDDED CREATIVE CONTEXTS BE REDUCED?

We hypothesise that most creative work is located in contexts with high complexity and tacit knowledge (i.e. Context D), the assumption being that the knowhow involved in the creative process tends to be tacit, and hence, difficult to codify. Most creative processes involve complex sequences of interactions between multiple skilled individuals, so both linkage ambiguity and characteristic ambiguity will be high. We suggest it may be possible to *reduce* both forms of causal ambiguity. Reducing ambiguity should help managers improve their ability to intervene in the creative process. But in spite of these reduced levels of ambiguity, compared to most other work, creative work will still generate higher levels of causal ambiguity.

In Figure 12.1, we set out four drivers of ambiguity: *tacitness*, *distance*, *complexity* and *time lag*. In order to reduce ambiguity in creative contexts and the resultant impacts on firm performance, we would therefore need to operate on these four drivers.

Reducing Tacitness

Creative work is reliant on the individual and collective knowhow of creative people. Generally, this is tacit knowhow, which resists codification. However, some aspects of tacit knowhow could be codified in the form of abstracted patterns of action or routines; these would be the *ostensive* aspect of routines identified by Martha Feldman (Feldman and Pentland, 2003). This higher-level abstraction from a specific performance of creative knowhow would be one step towards trying to capture some of the attributes, actions and dimensions of creative performance. This initial step in codifying creative work should provide some insights into the nature of creativity in this particular context.

This first step might then provoke insights into the capabilities, personal qualities, group interactions and other factors that are involved in creative work, which should then inform HR practices, particularly those involved with recruitment, training and team development.

Reducing Distance

It is difficult to comprehend creative work from a distance. HR professionals can reduce distance through closer involvement in creative work routines. Better still, if those with HR responsibilities were themselves experienced creatives, especially if their experience was within this particular firm. They would be likely to bring their intuitive understandings of these complex creative processes, which could then be augmented through the subsequent development of HR-specific managerial capabilities. Thus, the problem of distance is reduced as the insights into the specific creative context and judgements about appropriate HR interventions are integrated and embodied in the HR-trained creative.

Whilst generic HR practices may add value in contexts with low levels of ambiguity, they may well be damaging in creative work contexts. It might be appropriate in the management of creative work to allow creatives to engage in more "self-management". It might also be helpful to "buffer" creative activities, and preserve higher levels of autonomy. An important step here would be to develop creative employees so they can assume leadership roles.

In larger organisations, there may be benefits in embedding HR partners in these creative work contexts and encouraging them to engage with, interact with and observe the creative processes. In this way, they may be better able to enact changes that would be aligned with the peculiarities of creative work, as opposed to imposing "one size fits all" generic HR systems. For this approach to succeed, the judgements of these embedded HR professionals must be valued and respected by the central HR managers.

There would be value in creatives and others taking the time to share beliefs and generate collective understandings and insights into, for example, how the firm creates value and what causes success. Building *cause maps* might be one activity that could improve collective understandings and generate new insights and ideas into how to change the system (Ambrosini and Bowman, 2005; 2008).

Allowing creative staff more say in how they are developed and managed would also be beneficial. This could involve them being encouraged to challenge practices that do not fit the creative work context, and reinforce those that do. For example, the world-renowned Berlin Philharmonic Orchestra uses a secret ballot of its players to select their next conductor. Rather than, say, a governing board making this critical decision, it is delegated to the creative professionals, thus, reducing "distance". Reducing the separation between creative work and those who "manage" these creatives would have the effect of reducing distance and, in turn, reducing ambiguity.

Reducing Complexity

This may be the biggest challenge. Unpacking the actions and interactions involved in creative work, which is inherently complex, could prove to be counterproductive. Efforts to codify and simplify these complex processes may end up destroying the very qualities that make them valuable. Uncovering these "rules for riches" is an elusive endeavour, although many have tried. What makes for a blockbuster movie? Is there a formula for a hit single? Recently, the cinema-going public appears to be tiring of endless remakes, prequels and sequels of the original hit movie. These are clearly attempts at codifying and replicating the sources of success.

It may not be either feasible or desirable to try to simplify the socially complex creative process. But a shared understanding and a partial awareness of these complexities should prevent the introduction of potentially destructive and alien HR practices.

Reducing Time Lag

Attenuated creative processes generate linkage ambiguity. One way of reducing the impact of the time lag between the creative performance and realised value might be to try to track the longitudinal linkages in the whole process over the time of the project. As mentioned, cause mapping can be used to retrospectively analyse the causes of a successful project, which may have taken months or years to reach fruition (Ambrosini and Bowman, 2005; 2008). The flexible and loose structure of the cause map is well suited to unpacking complex attenuated value creation processes. Ideally, the map should be co-constructed by the team involved in the process. The map is a collectively produced and partial representation of the past. Creating maps for several projects enables some triangulation, and the process of building the maps inevitably generates insights and "aha!" moments. The collective learning involved in the map-building process can then be taken forward into subsequent projects. In attenuated creative processes, improving the timeliness and accuracy of feedback could reduce the damaging effects of long lead times. Mapping several past creative processes might suggest there are signifiers or warning signs along the way that could lead to more rapid responses in the unfolding creative journey.

THE FUTURE OF EMBEDDED CREATIVE WORK

Embedded creatives generate new knowledge that can add value to the core business. In Figure 12.2, we set out two contrasting value trajectories: A: new knowledge adds value over an extended time period; B: new knowledge adds value over a short time period.

Valuable new knowledge created by firms in the twentieth century would be likely to have a type A trajectory. This knowledge would be a strategic asset, one that is simultaneously valuable, rare, inimitable and non-substitutable (see Barney, 1991), and would provide a flow of sustained rents over years, if not decades. In contrast, the trajectory for new knowledge created in the twenty-first century would be more like a type B trajectory. The evidence for this would be the significantly reducing life cycle of firms. Whereas firms established in the twentieth century would have average life spans of decades, this life span has drastically shrunk in the twenty-first century. Reduced life spans are partly driven by the rapid diffusion of valuable knowledge across competing firms. The internet, cheap digital storage, social media, globalisation, a mobile workforce and isomorphic pressures all contribute to the diffusion of knowledge. Practices that were once unique sources of advantage for the firm that developed them rapidly become entry requirements for all firms.

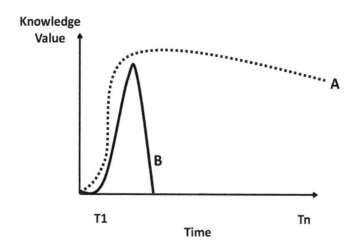

Figure 12.2   Contrasting value trajectories

If valuable employee knowhow can be partially captured through codification, then the knowledge shifts from an embodied form of capital to a separable form (Bowman and Swart, 2007). In this form, the knowledge becomes replicable and transferable. The ultimate form of codification is automation and artificial intelligence. In this context, it is important to consider the re-drawing of knowledge boundaries (Susskind and Susskind, 2015). That is to say, if valuable employee knowhow becomes codified (less tacit), that potentially frees up space for the embedded creative to engage in explorative, contextually sensitive creativity. This ultimately leads to the generation of new knowledge that can add value to the core business as well as the network within which the business is located. The future of embedded creative work therefore lies within the ability to generate this network value by delivering knowledge and outcomes for which other stakeholders, such as the client, do not have the expertise and/or experience to resolve.

There are two possible alternatives to consider here. In the first, human capital – that is, experience-rich knowledge and skills – *remains* at the heart of embedded creativity and artificial intelligence serves as an enablement to more efficient working. A central question here is which activities could be supported or replaced by technology to deliver solutions that are more efficient. However, this scenario reflects the current state of play, and future professional services work is likely to be subject to more disruptive and innovative adaptations of these technologies (Smets et al., 2017). In the second scenario, the introduction of increasingly capable systems, which reduces tacitness, complexity, time lag and distance, *replaces* the work of traditional embedded creatives. If this were to be the dominant scenario, then an important task for the future of embedded creative work is to identify which contexts and dimensions of creative skills will nevertheless continue to add value.

From Figure 12.2, in order for firms to sustain advantage, they need to continually create new and valuable knowledge. Ambidexterity refers to a firm's ability to continually create new knowledge (explore) whilst simultaneously exploiting existing knowledge (Turner et al., 2013). The challenge of ambidexterity is that the organisation structures, control systems, leadership styles, values and cultures appropriate to successful exploration are quite different from those required for the efficient exploitation of existing knowledge (Swart et al., 2019). One way of dealing with the conflicting requirements of exploitation and exploration is to separate these distinct activities. This can be done by establishing a separate marketing or research and development unit, and allowing it to develop organisational arrangements appropriate to nurturing exploration. Or exploratory work could be largely outsourced to creative firms, allowing the firm to focus almost exclusively on exploitation and efficiency (Tushman and O'Reilly, 1996). A more radical option, taken by Apple and Nike, is to outsource manufacturing entirely, thereby enabling the core business to focus on creativity.

But the short value life span of knowledge depicted in trajectory B probably now applies to firms in many industries. This means that in order to sustain competitive advantage, there is a requirement to continually develop new, valuable knowledge. Firms will become increasingly dependent on creative workers as the value lifecycle of existing knowledge reduces. The need for embedded creatives thus becomes an imperative for sustained advantage. The conclusion we must draw is that the effective stewardship of creative processes is now, and will become, an increasing critical challenge for managers.

It is also important to recognise that creatives embedded in industries outside the creative industries (Hearn and Bridgstock, 2014) may no longer be situated within a firm, but *operate across organisational boundaries, in networks* (Grimshaw and Rubery, 2005; Liftshitz-Assaf, 2017; Swart and Kinnie, 2013). This means that identification of the individual who generates *creative value that can be captured* becomes obscured. Furthermore, it is not only the nature of the boundaries within and across which embedded creative individuals operate, but it is also the property of their permanence. Take, for example, individuals who work in *temporary organisations* (Marchington et al., 2005) where cross-boundary creative value is generated, often intensely, for a short period of time before the network dissipates. The mix of insourcing and outsourcing of creative talent is also dynamic (Hearn and Bridgstock, 2014). These ways of organising creative work have also been characterised as liquid (Clegg and Baumeler, 2010). They are frequently found in the creative industries themselves. In this context, it is even more challenging to capture value generated from creative work at the level of the network. If we then return to Figure 12.2 and take curve B, it becomes clear that several individuals would be required to innovate over a period of time in order to continually generate value.

In essence, it is the combination of the key future drivers of the organisation of embedded creative work (i.e., cross-boundary and temporary work organisations) that points to the need to understand value generation and capture in liminal spaces. That is to say, embedded creatives increasingly operate "betwixt and between" other parties, most commonly employers and clients (Söderlund and Borg, 2017), in order to draw on their valuable tacit knowledge to innovate and generate value. The management of people in these liminal spaces (Cross and Swart, 2018) becomes challenging, given that organisations no longer have strategic freedom to unilaterally own, direct and control their embedded creatives. Furthermore, the identification of the source of creative talent and value generation is blurred and this means that it is more challenging to design key *people management* practices to attract, develop and retain creative individuals at both the organisation and the network levels.

This clearly holds implications for the management and unfolding of careers of embedded creatives. In the context outlined above, the focus of creativity shifts from the organisation to the individual. Importantly, creative work is also more networked. That is to say, embedded creatives are no longer tied to careers within organisations, but they have freedom to work across organisations, in networks. Interestingly, this tilts the bargaining power from the organisation to the individual, given that embodied human capital is network specific rather than firm specific. This has further implications for career management in so far as the focus shifts towards the increasing importance of social capital, which establishes the network within which the individual career is located. Embedded creatives therefore become focused on "who you know" (social capital), as well as cutting-edge creative skills, or "what you know" (human capital), in order to generate value.

CONCLUSION

Our argument centres on the problems of causal ambiguity in the management of creative work. We have focused particularly on circumstances where creative work is not the dominant task of the organisation; instead, it augments the core activities. However, the outcomes of creative work in these contexts may well be critical competitive advantages and if the dominant managerial approaches are geared to the core activity, there may well be detrimental impacts of these practices on creative work. By exploring different combinations of characteristic and linkage ambiguity, we were able to see some of the implications of misaligned practices and we concluded the chapter with some suggestions that could reduce these problems. Creative work is inherently complex; what we need are management attitudes and behaviours that acknowledge this reality.

NOTES

1.   School of Management, Cranfield University, cliff.bowman@cranfield.ac.uk.
2.   ORCID iD: 0000-0002-5365-0686.

REFERENCES

Ambrosini, V. and Bowman, C. (2005). Reducing causal ambiguity to facilitate strategic learning. *Management Learning*, **36**(4), 493–512.
Ambrosini, V. and Bowman, C. (2008). Surfacing tacit sources of success. *International Small Business Journal*, **26**(4), 403–31.
Ambrosini, V. and Bowman, C. (2010). The impact of causal ambiguity on competitive advantage and rent appropriation. *British Journal of Management*, **21**(4), 939–53.
Barney, J. (1991). Firm resources and sustained competitive advantage. *Journal of Management*, **17**(1), 99–120.
Boon, C., Eckardt, R., Lepak, D. and Boselie, P. (2018). Integrating strategic human capital and human resource management. *International Journal of Human Resource Management*, **29**(1), 34–67.
Bowman, C. and Raspin, P. (2019). *Why understanding real customer needs is the key to a successful innovation strategy*. Open Business Council, 13 February. Retrieved from https://www.openbusinesscouncil.org/why-understanding-real-customer -needs-is-the-key-to-a-successful-innovation-strategy/. (Last accessed 21 February 2020.)
Bowman, C. and Swart, J. (2007). Whose human capital? The challenge of value capture when capital is embedded. *Journal of Management Studies*, **44**(4), 488–505.
Campbell, D.J. (1988). Task complexity: A review and analysis. *Academy of Management Review*, **13**(1), 40–52.
Caves, R.E. (2000). *Creative industries: Contracts between art and commerce*. Cambridge, MA: Harvard University Press.
Clegg, S.R. and Baumeler, C. (2010). Essai: From iron cages to liquid modernity in organization analysis. *Organization Studies*, **31**(12), 1713–33.
Coff, R. (1999). When competitive advantage doesn't lead to performance: The resource-based view and stakeholder bargaining power. *Organization Science*, **10**(2), 119–213.
Cross, D. and Swart, J. (2018). Professional liminality: Independent consultants spanning professions. *Academy of Management Proceedings*, *USA*, **1**, 14500.
Dierickx, I. and Cool, K. (1989). Asset stock accumulation and sustainability of competitive advantage. *Management Science*, **35**(12), 1504–14.
Feldman, M.S. and Pentland, B.T. (2003). Reconceptualizing organizational routines as a source of flexibility and change. *Administrative Science Quarterly*, **48**(1), 94–118.
Gioia, D.A. and Chittipeddi, K. (1991). Sensemaking and sensegiving in strategic change initiation. *Strategic Management Journal*, **12**, 433–48.

Grimshaw, D. and Rubery, J. (2005). Inter-capital relations and the network organisation: Redefining the work and employment nexus. *Cambridge Journal of Economics*, **29**(6), 1027–51.

Hall, R. (1992). The strategic analysis of intangible resources. *Strategic Management Journal*, **13**(2), 135–44.

Hambrick, D.C. (2007). Upper echelons theory: An update. *Academy of Management Review*, **32**(2), 334–43.

Hambrick, D.C. and Mason, P.A. (1984). Upper echelons: The organization as a reflection of its top managers. *Academy of Management Review*, **9**(2), 193–206.

Harvey, D. (1989). *The condition of postmodernity* (14th edn). Oxford: Blackwell.

Haskell, J. and Westlake, S. (2018). *Capitalism without capital: The rise of the intangible economy*. Princeton, NJ: University of Princeton Press.

Hearn, G. and Bridgstock, R. (2014). The curious case of the embedded creative: Managing creative work outside the creative industries. In S. Cummings and C. Bilton (eds), *The Handbook of Creativity Management* (pp. 39–56). Cheltenham, UK, and Northampton, MA, USA: Edward Elgar Publishing.

Jameson, F. (1991). *Postmodernism, or, the cultural logic of late capitalism*. Durham, NC: Duke University Press.

King, A.W. (2007). Disentangling interfirm and intrafirm causal ambiguity: A conceptual model of causal ambiguity and sustainable competitive advantage. *Academy of Management Review*, **32**(1), 156–78.

King, A. and Zeithaml, C.P. (2001). Competencies and firm performance: Examining the causal ambiguity paradox. *Strategic Management Journal*, **22**(1), 75–99.

Kraaijenbrink, J., Spender, J-C. and Groen, A.J. (2010). The resource-based view: A review and assessment of its critiques. *Journal of Management*, **36**(1), 349–72.

Liftshitz-Assaf, H. (2017). Dismantling knowledge boundaries at NASA: The critical role of professional identity in open innovation. *Administrative Science Quarterly*, **63**(4), 746–82.

Lippman, S. and Rumelt, R. (1982). Uncertain imitability: An analysis of interfirm differences in efficiency under competition. *Bell Journal of Economics*, **13**(2), 418–38.

Marchington, M., Grimshaw, D., Rubery, J. and Willmott, H. (eds) (2005). *Fragmenting work*. Oxford: Oxford University Press.

Mintzberg, H. and Waters, J.A. (1985). Of strategies, deliberate and emergent. *Strategic Management Journal*, **6**(3), 257–72.

Reed, R. and De Fillippi, R.J. (1990). Causal ambiguity, barriers to imitation, and sustainable competitive advantage. *Academy of Management Review*, **15**(1), 88–102.

Smets, M., Morris, T., Von Nordenflycht, A. and Brock, D.M. (2017). 25 years since "P2": Taking stock and charting the future of professional firms. *Journal of Professions and Organization*, **4**(2), 91–111.

Smircich, L. and Stubbart, C. (1985). Strategic management in an enacted world. *Academy of Management Review*, **10**(4), 724–36.

Söderlund, J. and Borg, E. (2017). Liminality in management and organization studies: Process, position and place. *International Journal of Management Reviews*, **20**(4), 880–902.

Stoneman, P. (2010). *Soft innovation: Economics, product aesthetics, and the creative industries*. Oxford: Oxford University Press.

Susskind, R. and Susskind, D. (2015). *The future of the professions: How technology will transform the work of human experts*. Oxford: Oxford University Press.

Swart, J. and Kinnie, N. (2013). Managing multi-dimensional knowledge assets: HR configurations in professional service firms. *Human Resource Management Journal*, **23**(2), 160–79.

Swart, J., Turner, N., Van Rossenberg, Y. and Kinnie, N. (2019). Who does what in enabling ambidexterity? Individual actions and HRM practices. *International Journal of Human Resource Management*, **30**(4), 508–35.

Turner, N., Swart, J. and Maylor, H. (2013). Mechanisms for managing ambidexterity: A review and research agenda. *International Journal of Management Reviews*, **15**(3), 317–32.

Tushman, M.L. and O'Reilly, C.A. (1996). Ambidextrous organizations: Managing evolutionary and revolutionary change. *Harvard Business Review*, **38**(4), 8–30.

Walsh, J.P. and Ungson, G.R. (1991). Organizational memory. *Academy of Management Review*, **16**, 57–91.

Weick, K.E., Sutcliffe, K.M. and Obstfeld, D. (2005). Organizing and the process of sensemaking. *Organization Science*, **16**(4), 409–21.

PART IV

Educating for the future of creative work

13. Creativity 2.0: new approaches to creative economy work and education in the creative industries

Chris Bilton[1]

INTRODUCTION

Creativity research has repeatedly highlighted a distinction between two types of creative thinking process: divergent and convergent, adapters and innovators, left brain and right brain. This relates to a duality in the definition of *creativity*, poised between "novelty" and "value", and to different "stages" in the creative thinking process. In popular culture, we have tended to highlight original thinking and novelty over adaptive thinking and value, not least because the former appears to be more elusive and mysterious. Creativity is also, according to Boden (1990) and other creativity theorists, the element that distinguishes humans from machines, connected to the idea of human consciousness.

Yet adaptive thinking has an important part to play in creativity, especially in the creative industries themselves. Here, attention has shifted from the content of cultural products to the context of cultural consumption. Digital platforms have transformed the nature of cultural consumption, allowing more agency to consumers to "co-create" meaning and value. These technologies were identified as *Web 2.0*, a term describing apps and platforms that changed digital media from a broadcast model to a two-way flow based on collaboration and exchange. In this chapter, we consider a *Creativity 2.0* model of networked creativity adapted to our connected, digital culture. This model places greater emphasis on the latter stages in the creative process, on adaptation and user experience, rather than invention and product innovation. This more expansive definition of creativity also fits with this book's shift from *creative industries* to *creative economy* as the locus of creative work.

CREATIVITY AND THE CREATIVE ECONOMY

Creativity has long been recognised as a dualistic, multi-faceted entity that encompasses different ways of thinking (process), different combinations of groups and individuals (people) and different outcomes (product). Much of the critical literature on creativity has had a polemical edge, aimed at correcting popular assumptions about creativity as the preserve of gifted individuals, special talents and behaviours, a specialist task or "stage" in thinking, a pursuit of "mere novelty". There is broad acceptance that creativity requires multiple types of thinking (Gardner, 1984; Sternberg, 1988), a combination of different skills and talents (Kirton, 1984; Weisberg, 2010), a collaboration across creative teams, systems and organisations (Amabile, 1988; Csikszentmihalyi, 1988; Sawyer, 2006), and an outcome that is not only novel, but also valuable (Boden, 1994, pp. 75–9).

Despite this emphasis on a "sociocultural" paradigm of creativity (Sawyer, 2006), old hierarchies persist. Creative systems and teams still revolve around talented individuals. Theories of the creative process still privilege the moment of ideation over the painstaking processes of preparation and verification that precede and follow the flash of "illumination". Even attempts to de-essentialise and demystify creativity still reinforce a distinction between *creative* and *non-creative* roles (Glăveanu and Lubart, 2014).

This chapter argues that in today's creative economy it has become increasingly difficult to separate the creative idea from the expression and delivery of that idea, and this necessarily involves a *supporting cast* of co-workers and a suite of skills, techniques and technologies that are beyond the scope of the creative individual, no matter how talented. This mutual dependency is heightened by the increasing interdependence between specialised technologies of production and distribution in the creative industries.

Historically, the supporting cast has been positioned at a secondary order of importance and abilities beyond the creative individual. Administrative or "back-office" workers in the media and entertainment sectors are classified according to standard industrial and occupational codes (SIC and SOC[2] respectively) as pursuing a non-creative job in a creative industry. Within the organisation, this role separation has been entrenched by cultural and organisational hierarchies and divisions, which often result in mutual hostility and organisational dysfunction. Beyond the single organisation, the non-creative work of adapting, marketing and delivering creative products has been distributed unevenly along the value chain, which in turn reflects a fundamental divide between *content creators* and *gatekeepers* or *intermediaries* (Hirsch, 1972). Again, such divisions have been accompanied by mutual suspicion, if not outright hostility.

With the transformative and disruptive effects of digital technology on media and creative industries over the past 20 years, familiar divisions between production and consumption, between creation and mediation, and between art and technology, have become increasingly tenuous. The line between creative and non-creative roles is blurred also by the emergence of an *experience economy*, in which consumption and exchange around creative content are often as important – and creative – as the content itself. These processes of exchange and interaction are increasingly likely to be mediated through digital technologies.

As we transition from the always unsatisfactory definition of creative industries to the new rhetoric of creative economy, we may at the same time need to rethink what we mean by creativity. A creative economy is no longer restricted to certain "cultural", "media", "entertainment" or "audio-visual" sub-sectors such as music, TV, film or performing arts. Instead, we can recognise a much wider range of organisations and sectors that are loosely based on communicating meaning and value, from museums and theatres, to advertising, design, events management, user experience design or tourism. Technology companies play a central role in the experience economy, both in terms of experience design (using algorithms and consumer data to target individual preferences) and in terms of experience delivery (providing an extended and immersive experience beyond the core product) (Hesmondhalgh and Meier, 2018).

The interdependence between the art of production and the technologies of consumption in the creative economy forms the background to this chapter. First, the chapter will review the relationship between art and technology, based on the notion of technologies providing *affordances* for creativity. Then, the chapter will consider the implications of this digitally mediated creative economy for skills and training.

CREATIVITY, TECHNOLOGY AND AFFORDANCES

In today's creative economy, creativity has come to encompass more than good ideas. The packaging and experience of the idea are at least as valuable as the raw content. In this experience economy, digital technology plays a crucial role in corralling, retaining and commodifying consumer attention, to the point where it is difficult to separate products and services from the platforms and experiential frameworks that mediate them. This shift in values is reflected in the economic structure of the creative and media industries. Consumers expect cultural content to be free, yet they will pay for the technologies, platforms and channels that make this free content accessible, shareable, relevant or personal to them. Consequently, while content producers are competing in a saturated market with low margins, weak bargaining positions with consumers and intermediaries, and low to zero wages, the digital platforms that package this content (giving it away for free, but commodifying the data of those who consume it) have grown into some of the richest, fastest-growing companies on the planet. The Big Four of Apple, Facebook, Google and Amazon have, at various points, either approached or surpassed a market capitalisation of a trillion dollars and increasingly dominate the "creative" industries of music, publishing, film, TV and news. In China, the dominance of Tencent, Alibaba, Baidu and Sina-Weibo, along with businesses they own wholly or in part such as NetEase, iQiyi and WeChat, tells a similar story. Ideas are cheap, but the packaging and sharing of other people's ideas, identities and consumption habits is big business.

This restructuring of the creative economy necessitates a reassessment of both the nature of creativity itself, and the skills needed to succeed in a digitally mediated creative economy. For those in the business of creating content, the challenge is how to take control of the ways in which their content is delivered and experienced. This in return requires a re-orienting of what it means to be an artist in the twenty-first century. Musicians are musical entrepreneurs, authors are bloggers and self-publicists, and film-makers must think about the consumers sitting behind screens or watching through 3-D glasses, not just about stories and characters. Of course, creative industries have always combined creativity with business acumen. But the reliance on digital mediation requires something more, and changes the definition of creativity to encompass the ways in which ideas are adapted, shared and used. What was previously a secondary task, something entrusted to agents or marketers, has now become integral to story-telling. Digital technologists are no longer simply tasked with the administration of digital assets (Bennett, 2020) and must be taken seriously as part of the creative workforce. Digital platforms empower fans, not only to re-create or co-create their own version of a given piece of content (Jenkins et al., 2013), but also to enrich the value of the original work (Baym, 1998). Managing these information flows between author and reader, through digital platforms, has become integral to the creative value of cultural and media products.

The changing role of the artist is reflected in changing job descriptions for "creative" work. In the United Kingdom (UK), a recent report examined 35 million job advertisements and found a strong correlation between creative and digital skills. These *createch* skills were often associated with particular "creative" software, such as Adobe Photoshop. Of the occupations most reliant on createch skills, graphic design and photography unsurprisingly have creative and digital skills embedded within them; but createch skills were also found to be in demand among artists, art directors and producers (Bakhshi et al., 2019).

The idea that creative processes and products in the creative industries are mediated through technology is nothing new. Creative industries are the result of applying technologies of production and consumption to cultural artefacts, thereby destroying what Walter Benjamin (2008) called the *aura* of the art object and making the work of art available for mass consumption. According to the Frankfurt School, the process of mechanical reproduction is one of commodification, replacing authentic cultural experiences with a false promise. A similar scepticism surrounds today's mediated cultural products; however, it is based less on nostalgia for an unmediated, authentic product, and more on the dominant role of technology companies that have secured a monopoly on consumer data and consumer experiences (Keen, 2008; Lanier, 2010).

Despite this critique, it is clear that technology also plays an enabling role in cultural production. In the music industry, commentators including Simon Frith (1986) and Keith Negus (1992, p. 86) have noted that technology has made possible new forms of creativity and new styles of music. The opposition between purists and technologists in this sense is artificial. Yet, doubts remain about the limits of technology in the cultural sphere. Technology can *enable* human creativity, but could technology ever *replace* it? This is the question posed by Byron's daughter Lady Lovelace in 1842, known as the *Lovelace question*: Could a machine ever replicate human creativity, or is it limited to performing only those functions programmed into it by its (human) creators (Boden, 1990)?

One answer to this question is the notion of *affordances* (Glăveanu, 2012). An affordance can refer to any resource that makes an action possible, including, for example, finance, technologies, networks or human resources. On the other hand, an affordance can also restrict or channel action within a narrow framework of possibilities. So, on the one hand, creative technologies can enable or "liberate" creativity (Zagalo and Branco, 2015); however, on the other hand, they might also work against it. According to Jared Lanier's (2010) manifesto, technology, especially social media, removes agency and control, favouring "flatness in cultural expression" (p. 120).

Behind this discussion is a question about the relationship between freedom and constraint in the creative process. Contrary to the assumption that creativity equates to absolute freedom of expression, which allows individual creativity to be "unleashed", a majority of creativity theorists argue that creativity depends upon the constraints imposed by expertise, genre, technique or tradition. For Weisberg (2010), domain-specific expertise gives shape and direction to individual creativity. Csikszentmihalyi (1988) describes individual creativity framed by a creative system that comprises both a domain (a production culture) and a field (institutional channels and relationships). Margaret Boden (1994) likewise describes creativity taking place within a bounded conceptual space (pp. 79–84), testing and stretching the boundaries to eventual breaking point. The transformative power of creativity, according to Boden (1994), results not from "thinking outside the box", but thinking at the extreme inside edges of the box. Applying these notions of freedom and constraint to creativity and technology, the affordance of digital technology both enables and constrains (Moeran, 2014, pp. 41–2). This is not so much a contradiction, but more of a paradox: constraints enable creativity.

Furthermore, while affordances for creativity exist, they must also be perceived, recognised and utilised (Glăveanu, 2012). This requires some broader *meta-level awareness*, understanding what the affordance (technology) can and cannot enable, and turning it to account. For an example of how this might work in practice, we might consider experimental attempts to answer the Lovelace question using technology to replicate or replace human creativity. Hennig-Thurau and Houston (2019) describe two such experiments (pp. 306–10). The first was an attempt to compose a hit musical using algorithms to generate the plot and the music (the book and lyrics were man-made), resulting in the musical theatre show *Beyond the Fence*, which premiered at London's Arts Theatre in 2016. The second was an attempt over several years in the early 2000s by composer David Cope to use computing power to input the work of great composers, from Bach to Mozart, into an algorithm that then reproduced plausible versions of similar compositions, potentially on a massive scale.

Hennig-Thurau and Houston (2019) write persuasively on the power of algorithms to shape decision-making in the creative industries, rather than relying on mere "intuition" to green-light a creative project. Today's digital technologies are capable of analysing vast amounts of consumer data to predict consumer tastes, and streaming services such as Netflix and Spotify have, according to Hennig-Thurau and Houston, successfully incorporated this data into their programming decisions. However, when technology is used not only to predict consumer taste but also to create the artistic product, the results are more mixed. *Beyond the Fence* was not a hit with audiences or critics and had a limited run. One reviewer described the work as "bland, inoffensive, and pleasant as a warm milky drink" (Gardner, 2016, para. 1) as well as "risibly stereotypical" (Gardner, 2016, para. 5). Cope (1991) has continued to refine his project of computer-generated music. But he too acknowledges its limitations. The computer is incapable of making decisions or of recognising the value of its own output. Cope himself has the task of selecting and refining this output. In the end, as Lovelace predicted, the algorithm remains a tool, just as a keyboard or a guitar pedal are tools, enablers rather than surrogates. The key difference may be that Cope positions himself as an affordance for the composition algorithm rather than the other way around.

Hennig-Thurau and Houston (2019) are interested in the application of data to shape creative decisions. Yet, in this instance, while the individual decisions (Which note, in which order? Which plot element?) are plausible, the algorithmic process lacks a wider view of the composition process that allows artists not only to generate plots, but also to tell a story. As Cope (1991) discovered, the machine is not capable of recognising the quality of its own ideas, of discriminating between a good or bad innovation. We might call this missing ingredient emotional intelligence or self-awareness. With the right data input, an algorithm can generate endless variations on that input, resulting in plausible replicas of a Bach chorale or generic plots for a musical. But the machine is not (yet) capable of selecting which promising ideas are worth pursuing and stringing them together. In terms of creativity, the algorithm is better at idea generation than idea recognition. In the iterative loop of the creative process, this is a severe limitation because idea recognition is necessary to generate the next idea and build a sequence of ideas into a coherent narrative.

If technology is an affordance, it provides some capabilities that might accelerate or enhance a creative process; it might also, as Lanier (2010) suggests, be reductive, "flattening" human creativity. In creativity terms, affordances might result in what Boden (1990) terms *mere novelty* – and novelty without value (a subjective judgement) is not the same thing as creativity. Human intelligence is necessary to connect innovative elements into valuable outcomes and to organise the different stages or components of creativity into a coherent whole. Creative thinking and digital tools (including the "algorithmic creativity" explored by Cope) can be seen as complementary, but they need this meta-level cognitive connection to turn affordances into actions. This becomes even more important in popular culture because fans' responses and fan creativity can, with the right digital architecture and the complicity of the copyright owner, be orchestrated and integrated to add value to "original" work. Digital fandom might therefore be classified as another affordance for creativity, provided that it can be integrated and recognised by the original creator.

CREATIVE COMPETENCES: THE PROBLEM OF SPECIALISATION

Whereas affordances describe externalities that frame the creative process, *competences* describe capabilities within creative individuals or teams. Like external affordances, internal competences can both constrain or enable action. We know that creative individuals and creative teams must encompass a range of different thinking styles and cognitive processes, from divergent to convergent, from diligent perseverance to spontaneous risk-taking, from rational to intuitive. In the first part of this chapter, this multiplicity was linked to the increasing reliance on multi-talented teams to develop and deliver creative content in the digital creative economy. As with affordances, competences are not in themselves intrinsically "creative" and the elements that comprise creative cognition are less important than the connections that join them. This in turn requires some meta-level awareness or governance that can discriminate and recalibrate between often opposing tendencies.

There have been numerous attempts to categorise different cognitive elements within creative teams, or within teams in general. Belbin's (1993) *team roles* is one of the more commonly applied frameworks; De Bono's *Six thinking hats* is another (De Bono, 1986). One limitation of these taxonomies is that first they can appear to privilege one type of thinking over another (Is the "plant" more creative than the "completer-finisher"?). Another weakness is that they can trap individuals in rather stereotypical, limited roles that do not allow them to change, develop or rediscover themselves (McCrimmon, 1995). In a creative team especially, such fixity might be particularly deadening.

The tendency to type-cast individuals in teams is particularly pronounced in the so-called creative industries themselves. Without wanting to dwell too long on the definition, creative industries deploy human imagination and intelligence to devise *symbolic goods*: products or services whose primary purpose and value lie in their ability to communicate meaning. For example, film, music and advertising all fit this definition. These industries are usually project-based, assembling creative teams for the duration of a project to con-tribute specialist skills for a limited period of time. At the end of the project, the team dissolves, only to reassemble in new configurations around another project. Team members are valued for their specialist contribution, and their ability to pitch for work depends upon freelance individuals and specialist firms highlighting their talents as distinctive, scarce or even unique. Consequently, the creative industries, particularly those involved in the generation of content as opposed to exploitation or delivery, are typically dominated by small, tem-porary project-based enterprises collaborating in networks. Depending on the sector, most "organisations" in the creative industries number fewer than 10 employees, with other individuals joining ad hoc according to the demands of the project in hand. The characteristic mode of production is post-Fordist *just in time* delivery by networks of specialists collaborating in temporary teams.

With the growing reliance on digital technologies for both the dissemina-tion and production of cultural products, any creative team in the creative industries is likely to include its share of technology specialists alongside creative specialists. From music technology to computer-generated imagery to computer-aided design, specialist technologies of production in the music, film or design industries require specialist talents to operate them. Creative technologists thus form a significant proportion of the network of specialists that lie behind the creative team.

A weakness in this project ecology is the absence of any system of gov-ernance that can ensure continuity, allow for reflection and self-awareness or gather and archive collective memory (Grabher, 2004). In the absence of any permanent core or centralising hub, what is to stop these teams of specialists collapsing into pointless repetition, or self-destructive and self-defeating behaviours? And what is there to ensure that such dissimilar, even opposing, mindsets can collaborate effectively?

In reality, people working in the creative industries are used to collaborating and adapting their behaviours to accommodate differences of opinion, ideology or attitude. Against the risk of over-specialisation and fragmentation, there is an opposing tendency to self-assess, adapt and self-reflect. Contrary to the stereotype of "difficult" creative individuals (who surely do exist!) anybody seeking a creative career quickly learns to sacrifice individual goals and preferences to the needs of the shared project. This in turn requires a degree of self-awareness, self-restraint and adaptability. While individuals and micro-businesses may emphasise competitive differences when pitching for contracts and projects, they rely upon cooperative abilities to deliver them.

When assessing the competences required for a creative team, it is possible to make a case for any number of specialist attributes or talents. As soon as these are compiled into a list, the apparent contradictions and tensions between these different competences become glaringly obvious. Perhaps more important than any technical or artistic ability, a key attribute of any member of the creative team is their awareness of their own and others' abilities, of how these can dovetail together and of how individual competences mesh with the overall aims of the project. This "meta-awareness" reflects the ambidextrous, "bisociative" nature of creativity, described by psychologists as "tolerance for contradictions" (Barron, 1958), "constructive, synthesizing, unifying and integrative" (Maslow, 1987, p. 162) or "multifaceted" (Sternberg, 1988), and by the novelist F. Scott Fitzgerald in his 1936 essay "The Crack-up" as "the ability to hold two opposed ideas in the mind at the same time, and still retain the ability to function" (Fitzgerald, 2005, p. 139).

If creative cognition requires an ability to bridge different competences, attitudes and thinking styles, how is this to be cultivated through education? Creative skills are individualised and specialised; the same could be said of technological skills. The next section considers how this risk of over-specialisation and talent "silos" can be addressed in the curriculum.

TECHNOLOGY AND CREATIVITY: THE EDUCATIONAL CHALLENGE

Bridging creative, entrepreneurial and technological disciplines places strains on an education system that has tended to treat these as separate subjects, taught through different curricula and institutions.

In the creative industries, stereotypical assumptions about creativity as self-expression and freedom led to the downgrading of "humdrum" work (Caves, 2002) and a hierarchical separation between creative and *uncreative* work in organisations (Bilton, 2015). In the creative economy, the need for collaboration and multi-faceted creative teams moves in the opposite direction towards greater convergence. By and large, our education systems have not adapted to deal with this convergence. One obstacle is an outdated definition of creativity as an "artistic" specialism, rather than something that cuts across the entire curriculum.

In the report commissioned by the UK government to examine the position of creativity in English schools, Ken Robinson called for a "systemic strategy" that recognises creative and cultural education as "general functions of education", not "subjects in the curriculum" (NACCCE, 1999, p. 6). The report emphasises that creativity is not the preserve of "the arts" or "the creative industries" (p. 28), nor is it restricted to exceptional individuals. Rather, there are "a wide range of intelligences" (including emotional and intellectual varieties) that can all have value in the creative process (pp. 38–9); the creative process is accordingly "multidimensional" (p. 41). Addressing how "creative and cultural education" can be delivered, the report highlights the need for "a balance in the curriculum" between arts and sciences and technology, against the "assumed hierarchy" between core and foundation subjects (p. 59). It is also suggested (p. 62) that a focus on the arts and humanities will be a means of understanding and engaging with the transforming power of *new* technologies ("new" here being what we might refer to as "digital").

Robinson's report (NACCCE, 1999) contains recommendations for funding and teaching training, in addition to detailed recommendations for restructuring the school curriculum, including:

- abolishing the hierarchy between core and foundation subjects (which has tended to prioritise or "protect" science subjects);
- reorganising the curriculum around learning areas or subject groupings rather than discrete subjects;
- moving towards greater autonomy, choice and "self-directed learning";
- resisting a growing emphasis on summative assessment ("teaching to the test").

Twenty years on, none of these recommendations has been implemented. Indeed, more recent educational reforms in the UK have hardened the distinction between STEM subjects (science, technology, engineering, mathematics) and humanities. The newly minted English Baccalaureate (EBacc), designed as the new gold standard to measure school and student performance at secondary level, does not require any creative arts subjects. The polarisation of STEM subjects and creative arts subjects begins at secondary level and transfers into universities. In 2011, the Warwick Commission found that "only 8.4% of students accepted for Creative Arts & Design undergraduate courses had taken Maths A-level, and only 5% of those accepted for Maths and Computer Science courses had studied A-level Art and Design" (Neelands et al., 2015, p. 45). At the same time, the space for cross-curricular formative education has shrunk (partly due to cuts in local school budgets, but also in response to government education reforms that encourage an exclusive focus on "core" – principally STEM – subjects). The pressure on young people to pursue a specialised education at secondary school is further exacerbated at tertiary level by the introduction of student fees for higher education, encouraging students to view education in terms of commercial employability rather than other forms of development (social, personal). Universities also play their part in this narrowing of "creative" horizons.

While Robinson's report (NACCCE, 1999) has been widely praised for its ambition, this may also have been its undoing. Perhaps, in commissioning such a report in the first place, the UK government anticipated a narrowly defined examination of arts provision in schools. Instead, in pursuit of a *creative* curriculum, the report ranged across every aspect of school education, setting challenging targets for a more integrated, balanced and inclusive curriculum. Did Robinson go beyond his brief? According to the arguments in this chapter, such a wide-ranging approach is inevitable to advance an integrated, creative education; however, from the perspective of governments that seek quick wins and are reluctant to undertake expensive, risky and electorally contentious reforms, it is perhaps unsurprising that such an approach has proved politically unpalatable.

In terms of its highly specialised, subject-based approach to secondary and tertiary education, the English system may be an outlier. As pointed out in the document, many of the reforms advocated in Robinson's report were already common practice in other countries (NACCCE, 1999). But in the gap between recommendations and implementation, Robinson's report highlights the wider difficulties in promoting the kind of balanced, integrated and multi-faceted education necessary for the future of work in the creative economy. A further explanation for the "failure" of the Robinson report may be the argument that the UK government preferred to pursue its own "dominant pro-market construction of creativity" (Neelands and Choe, 2010, p. 301) rather than Robinson's pervasive, democratising vision of creativity as socially and personally transformative. Here, too, the UK's political and economic pragmatism may be part of a wider challenge for creative education, not just a problem in English schools. Where the UK fails to lead, others have followed. In their examination of Singapore's investment in creative education in universities, Comunian and Ooi (2016) note that initiatives appear to be economically driven, benefiting a relatively small proportion of students rather than developing "a local ecosystem of creative and cultural production" – and that Singapore's "hierarchy of competences" continues to rank arts and culture behind scientific knowledge (pp. 74–5). Singapore's ambitious programme of educational reform, designed to promote employment in the creative economy, appears to entail greater specialisation and individualisation in education, not less.

Perhaps, then, if creativity rests upon "a synergistic interaction between science and technology on the one hand and the arts and humanities on the other" (NACCCE, 1999, p. 76), the solution is not going to be found in schools. Short of the kind of wholesale reforms promoted by the Robinson report, formal state education systems are framed by a range of political and economic priorities that mitigate against the free-flowing, synergistic approach that might release and transform collaborative creativity. For technologists and artists to collaborate, this chapter has advocated a meta-level awareness of both one's own and others' "creativities", allowing individuals with different backgrounds, skills and experiences to collaborate across specialisms. This is the essence of Creativity 2.0, and it is unlikely to be found in a formally assessed school curriculum. It might be found in extra-curricular activities, non-assessed personal projects, and visits and partnerships beyond the school gates (all of which are being squeezed out by a combination of tightening budgets and narrowing targets, at least in the state-supported sector). More likely, it will be found outside our schools and universities entirely.

Just as Robinson's report (NACCCE, 1999) extended its scope beyond "arts" subjects across the curriculum and into partnerships and organisations outside the school, we may need to expand our definition of creative education still further. The kind of formative interactions at the core of Creativity 2.0 take place in workplaces and communities, in addition to formal educational settings. Providing space and time for these encounters to occur then becomes a task for organisational managers and urban planners, as well as freelancers and micro-enterprises, especially those working in the digital creative economy. Cultivating a meta-level awareness of connections between disciplines and skills is something learned from experience, through practice. Allowing time for curiosity, risk and experimentation in the workplace, allowing greater social connection in mixed developments, and designing flexible workspaces that encourage interaction and collaboration can all play a part in providing these opportunities for mutual self-discovery and awareness. All of these things in turn require a tolerance for waste, risk and failure, rather than a targeted, narrowly focused approach to solving immediate problems and tasks, which is not easy in a time of rising competition, declining infrastructure and reduced state funding.

CONCLUSION

This chapter has argued for an expansive definition of creativity that includes the delivery and experience of creative content, not just its origination. This seems to resonate with an expansive and "contextual" definition of innovation in the creative economy. According to Wijngaarden et al. (2019), creative workers considered innovation not as a goal but as a by-product of the creative process, measured in terms of artistic or social purpose, not just as the generation of something new. This means acknowledging that some of those processes, people or interventions that might previously have been regarded as uncreative or oppositional to the creative process are in fact integral to it. Uncreative people can add value by recognising and filtering creative ideas or by highlighting the creativity of incremental change, innovating by "thinking *inside* the box". Above all, uncreativity connects innovation to value, integrating what is new with the affordances that make what is new achievable and recognisable to others, locating novelty in a field of value.

In today's creative economy, digital technology has become one of the most critical of these affordances for creative value. At the level of distribution and dissemination, digital platforms connect creative ideas to audience experiences, and feed information about those audience experiences back into the creative process. At the level of production, technologies provide accessible and affordable means to democratise the creative process, allowing professional and amateur creatives access to the means of production. Web 2.0 provided digital tools for users to generate their own content, and Creativity 2.0 describes this merging of creative ideas with technological means.

Creativity 2.0 depends upon a merging of artistic and technological competences, both at the individual and collective level. For the individual, some meta-level awareness is needed to recognise the strengths and limitations of a singular approach to creative work and to see past the raw idea (especially if that idea is one of our own). At a collective level, creative teams require not only a diversity of talents and approaches, but also the emotional intelligence to bridge different, sometimes opposing, visions and mindsets. Diversity is a necessity in a creative team, but diversity must also be recognised and acted upon.

Training people to work in collaborative creative teams will not be possible if education is understood and measured simply in terms of individuals acquiring specialist skills. Digital technology offers significant affordances for creative work. The competences needed to act upon these affordances can be nurtured through education. First, education at secondary and tertiary level needs to allow students to bridge between creative arts subjects and science and technology subjects. Second, creative and cultural education should be directed not only towards skills development, but also towards an understanding of creative work as a collective practice, not (or not only) an individual talent. Allowing and encouraging students from different disciplines to work together will help them to develop this meta-level appreciation of diversity and combined talents through students' own experiences. Third, creative education needs to be embedded in a participatory educational culture that nurtures localised ecosystems of collaboration, not just the unleashing of individual talents. As noted by Robinson and his colleagues (NACCCE, 1999), this approach to creative and cultural education necessarily extends into partnerships and projects outside the educational institution and beyond the exam-based curriculum.

This chapter has considered the need for a more integrated approach to education, extending "creative education" beyond its association with arts and humanities, and applying principles of creative thinking across every aspect of the curriculum. The failure to implement most of the recommendations of the NACCCE (1999) report on cultural and creative education in the UK highlights the political and pragmatic challenges of pursuing this approach in our increasingly goal-oriented education systems. As the example of Singapore indicates, the UK emphasis on specialisation and hierarchical distinctions between arts and sciences is hardly unique. The combination of creative and digital skills described in this chapter is symptomatic of the more collaborative and interdisciplinary practices emerging in the new creative economy. In a special issue examining creative careers and higher education in the UK and Australia, Bridgstock et al. (2015) argue that preparing students for creative careers calls for a more holistic approach, specifically "recontextualisation and reinterpretation" of previously acquired skills, knowledge and practices (p. 340). It is clear that preparing for careers in our future creative economy is not only a task for schools and universities: we need also to look beyond formal education towards the way we organise our workplaces and our working routines, and the way we plan our urban spaces. Through a better understanding of these challenges, we can hope that even if education cannot provide the solution, it can at least cease to be part of the problem.

NOTES

1. Centre for Cultural & Media Policy Studies, School of Creative Arts, Performance and Visual Cultures, University of Warwick, c.bilton@warwick.ac.uk.
2. These codes are used in the UK to classify jobs and occupations in UK census data. Standard Industrial Classification (SIC) classifies businesses according to the economic activity they are engaged in. Standard Occupational Classification (SOC) classifies individual workers according to the work they do and the skill

level required. Historically, it has been difficult to apply existing categories to creative industries and creative work respectively, making it difficult to quantify the scope of the sector. An improved classification system was introduced in 2014, promising a more robust measurement of the UK's creative economy. There are similarities here with the Creative Trident model in Australia.

REFERENCES

Amabile, T.M. (1988). A model of creativity and innovation in organizations. *Research in Organizational Behaviour*, **10**, 123–67.

Bakhshi, H., Djumalieva, J. and Easton, E. (2019). *The creative digital skills revolution*. London: Creative Industries Policy Evidence Centre / NESTA.

Barron, F. (1958). The psychology of imagination. *Scientific American*, **199**(3), 251–66.

Baym, N. (1998). Talking about soaps: Communication practice in a computer-mediated culture. In C. Harris and A. Alexander (eds), *Theorising fandom: Fans, subculture and identity* (pp. 111–29). New York, NY: Hampton Press.

Belbin, R.M. (1993). *Team roles at work*. Oxford: Butterworth-Heinemann.

Benjamin, W. (2008). *The work of art in the age of mechanical reproduction* (J.A. Underwood, trans.). London: Penguin. (Originally published 1935.)

Bennett, T. (2020). Towards "embedded non-creative work"? Administration, digitisation and the recorded music industry. *International Journal of Cultural Policy*, **26**(2), 223–38.

Bilton, C. (2015). Uncreativity: The shadow side of creativity. *International Journal of Cultural Policy*, **21**(2), 153–67.

Boden, M. (1990). *The creative mind: Myths and mechanisms*. London: Weidenfeld and Nicolson.

Boden, M. (1994). What is creativity? In M. Boden (ed.), *Dimensions of creativity* (pp. 75–117). Cambridge, MA: MIT Press.

Bridgstock, R., Goldsmith, B., Rodgers, J. and Hearn, G. (2015). Creative graduate pathways within and beyond the creative industries. *Journal of Education and Work*, **28**(4), 333–45.

Caves, R.E. (2002). *Creative industries: Contracts between arts and commerce*. Cambridge, MA: Harvard University Press.

Comunian, R. and Ooi, C.S. (2016). Global aspirations and local talent: The development of creative higher education in Singapore. *International Journal of Cultural Policy*, **22**(1), 58–79.

Cope, D. (1991). *Computers and musical style*. Oxford: Oxford University Press.

Csikszentmihalyi, M. (1988). Society, culture, and person: A systems view of creativity. In R.J. Sternberg (ed.), *The nature of creativity: Contemporary psychological perspectives* (pp. 325–39). Cambridge: Cambridge University Press.

De Bono, E. (1986). *Six thinking hats*. London: Penguin.

Fitzgerald, F.S. (2005). The crack-up. In F.S. Fitzgerald and J.L. West III, *My lost city: Personal essays, 1920–1940*. Cambridge: Cambridge University Press. (Essay originally published 1936.)

Frith, S. (1986). Art versus technology: The strange case of popular music. *Media, Culture & Society*, **8**(3), 263–79.

Gardner, H. (1984). *Frames of mind: The theory of multiple intelligences*. London: Heinemann.

Gardner, L. (2016). Beyond the fence review: Computer-created show is sweetly bland. *The Guardian*, 29 February. Retrieved from https://www.theguardian.com/stage/2016/feb/28/beyond-the-fence-review-computer-created-musical-arts-theatre-london.

Glăveanu, V.P. (2012). What can be done with an egg? Creativity, material objects and the theory of affordances. *The Journal of Creative Behavior*, **46**, 192–208.

Glăveanu, V.P. and Lubart, T. (2014). Decentring the creative self. *Creativity and Innovation Management*, **23**, 29–43.

Grabher, G. (2004). Learning in projects, remembering in networks? Communality, sociality and connectivity in project ecologies. *European Urban and Regional Studies*, **11**(2), 103–23.

Hennig-Thurau, T. and Houston, M. (2019). *Entertainment science*. Cham: Springer International.

Hesmondhalgh, D. and Meier, L.M. (2018). What the digitalisation of music tells us about capitalism, culture and the power of the information technology sector. *Information, Communication & Society*, **21**(11), 1555–70.

Hirsch, P.M. (1972). Processing fads and fashions: An organisation-set analysis of cultural industry systems. *American Journal of Sociology*, **77**(4), 639–59.

Jenkins, H., Ford, S. and Green, J. (2013). *Spreadable media: Creating value and meaning in a networked culture*. New York, NY: New York University Press.

Keen, A. (2008). *The cult of the amateur: How blogs, MySpace, YouTube, and the rest of today's user-generated media are destroying our economy, our culture, and our values*. New York, NY: Doubleday.

Kirton, M. (1984). Adapters and innovators: Why new initiatives get blocked. *Long Range Planning*, **17**(2), 137–43.

Lanier, J. (2010). *You are not a gadget: A manifesto*. New York, NY: Alfred A. Knopf.

Maslow, A. (1987). *Motivation and personality* (3rd edn). New York, NY: Harper & Row. (Originally published 1954.)

McCrimmon, M. (1995). Teams without roles: Empowering teams for greater creativity. *Journal of Management Development*, **14**(6), 35–41.

Moeran, B. (2014). *The business of creativity: Toward an anthropology of worth*. Walnut Creek, CA: Left Coast Press.

NACCCE (1999). *All our futures: Creativity, culture and education*. Report to the Secretary of State for Education and Employment and the Secretary of State for Culture, Media and Sport, May. Retrieved from http://sirkenrobinson.com/pdf/allourfutures.pdf.

Neelands, J. and Choe, B. (2010). The English model of creativity: Cultural politics of an idea. *International Journal of Cultural Policy*, **16**(3), 287–304.

Neelands, J., Belfiore, E., Firth, C., Hart, N., Perrin, L., Brock, S., Holdaway, D. et al. (2015). Enriching Britain: Culture, creativity and growth (The 2015 Report by the Warwick Commission on the Future of Cultural Value). Coventry: University of Warwick.

Negus, K. (1992). *Producing pop: Culture and conflict in the popular music industry*. London: Routledge.

Sawyer, R.K. (2006). *Explaining creativity: The science of human innovation*. Oxford: Oxford University Press.

Sternberg, R. (1988). A three-facet model of creativity. In R.J. Sternberg (ed.), *The nature of creativity: Contemporary psychological perspectives* (pp. 125–47). Cambridge: Cambridge University Press.

Weisberg, R. (2010). The study of creativity: From genius to cognitive science. *International Journal of Cultural Policy*, **16**(3), 235–53.

Wijngaarden, Y., Hitters, E. and Bhansing, P.V. (2019). "Innovation is a dirty word": Contesting innovation in the creative industries. *International Journal of Cultural Policy*, **25**(3), 392–405.

Zagalo, N. and Branco, P. (2015). The creative revolution that is changing the world. In N. Zagalo and P. Branco (eds), *Creativity in the digital age* (pp. 3–15). London: Springer.e

14. When dancers learn to teach dance: how creatives acquire expertise in multiple domains to improve employability[1]

Jose Hilario Pereira Rodrigues[2]

INTRODUCTION

Hearn and McCutcheon (2020) suggest that innovations in the creative economy, and the skills required to achieve them, are synergistic, drawing knowledge from four paradigms: "STEM (i.e., science, engineering and technology), business (i.e., economics, finance, entrepreneurship and management), social sciences (sociology, psychology, policy, law and social work) and MAD (media, arts and design)" (p. 22). Hearn and McCutcheon (2020) define creative knowledge as "replicative or novel aesthetic and/or expressive knowledge either separately or in synthesis with other forms of knowledge" (p. 17).

Bridgstock and Goldsmith (2016) show that creative work in education is a significant component of the creative workforce who work in sectors outside of the arts, that is, the *teaching artist* (Booth, 2003; Huddy and Stevens, 2011). Bridgstock (2015) further argues that future creative employability involves developing the *key-shaped individual*, that is, individuals with deep disciplinary expertise in one creative domain, connected to additional expertise in multiple alternative domains. Multidisciplinary expertise should improve employability for creative workers in the future. How such expertise is acquired is therefore a key question that this chapter addresses. A plurality of perspectives in relation to how expertise and experts should be defined, across many fields, is found within the literature on expertise (Baker et al., 2015; Cooke, 1992; Swann et al., 2015). In this chapter, I discuss how the innovative *ecological dynamics* (ED) *constraints-led approach* (CLA), can potentially add to creative knowledge expertise. This is illustrated with a study

by Rodrigues (2017), based on the ED-CLA, which investigated how expert
dance teachers acquired their pedagogical expertise. From this, five broad
learning design principles are formulated to suggest how the acquisition of
expertise, in a different knowledge paradigm, can be achieved to enhance
creative careers:

- encouraging formal and informal interaction with mentors and peers;
- engaging a variety of learning environments, with different task constraints, to allow natural evolution of expertise;
- engaging in dynamic environments to learn to deal with the unexpected and
  search for effective solutions;
- engaging in activities in which individuals can explore their own intrinsic
  dynamics and goal-setting;
- avoiding the pursuit of a "one right answer" ideal expertise behavioural
  model.

ECOLOGICAL DYNAMICS

The ED approach is strongly informed by the work of J.J. Gibson (1986) in the
field of ecological psychology. According to Gibson, the mutuality between
the individual and the environment emerges from the coupling between
mechanisms of perception and action, regulated by specific information, or
affordances (i.e., opportunities for action), generated by individuals' actions in
situated environments. Perception and action are inter-dependent and cyclical:
by perceiving environmental information an individual acts; and by acting, an
individual perceives further information (Gibson, 1986).

Critical to this understanding is the idea that action modifies the type
of affordances available to the individual (Withagen et al., 2012). Action
emphasises adaptive capabilities that have origin in the individual–
environment dyad, which permits individual coordination and behaviour
control to emerge (Gibson, 1986). It is important, therefore, to consider the
coupling between perception and action because it provides a foundation
for explaining behaviour in situated environments, while an individual is
trying to achieve a specific goal. For example, a dance teacher in a situated
environment picks up affordances in relation to functionalities that he or
she matches with her or his action capabilities; hence, there is a degree of
fit between environmental properties and the dance teacher's capabilities
for action.

ED conceptualises the person and their environment as a *dynamic system*
and considers each human as a whole, although they are composed of distinct
parts that connect with each other (Araújo et al., 2006). Changes to one part
might influence other parts; therefore, it is important to understand how each

part is unified and can affect other parts to understand human behaviour (Davids et al., 2008). Dynamic systems explain how the individual is capable of exploring the environment in order to develop functional patterns, while achieving a specific goal (Davids et al., 2008). The key features of dynamic systems include nonlinear behaviour (characterised by stable and unstable patterns), the potential for subsystems to influence other subsystems, multiple and varied degrees of freedom, and several levels of existence (Kauffman, 1993). For example, complex behaviour, such as teaching behaviour (Berliner, 2001; Shulman and Shulman, 2004), can be explained by the emergence of stable and unstable patterns of behaviour evolving from ongoing interaction in specific environments.

CONSTRAINTS-LED APPROACH

The analysis of the developmental pathway of expert dance teachers, within ED-CLA, posits functional interdependency among the individual, task and environment to explain the emergence of behaviour (Araújo et al., 2004; Davids and Araújo, 2010). Expertise acquisition involves learners' activities within specific environments (Araújo et al., 2009b); thus, it is dynamic and contextual in nature (Barab and Kirshner, 2001). The learning process can be mediated by understanding key environmental, task and individual constraints, potentially acting on individuals throughout their search for appropriate task solutions (Davids et al., 1999). Individuals perceive key information from these constraints, which operate as system control parameters, in order to regulate their action during goal-directed behaviour (Davids et al., 1999).

A thorough investigation of expertise includes the identification of the intrinsic dynamics of individuals, that is, the favoured behavioural predispositions that emerge from the interaction of environmental, task and individual constraints (Kelso, 1991) and the constraints that shape their behaviour (Phillips et al., 2010).

Rodrigues (2017) investigated the developmental experiences of dance teachers' acquisition of pedagogical expertise, using the CLA (Davids et al., 2008). A mixture of convenience and purposive sampling (Patton, 2002) was used to select and recruit 10 expert dance teachers of ballet and contemporary dance from an Australian tertiary institution (with a focus on vocational dance training). Specific criteria were applied to classify dance-teaching expertise. Participants were individually interviewed using a semi-structured, in-depth interview protocol to explore meaning and gain a holistic understanding of participants' professional life experiences.

Rodrigues (2017) identified five constraints influencing dance teachers' acquisition of pedagogical expertise. These constraints were divided into three categories, in accordance with the framework: environmental (e.g., mentors, role models, and students); task (e.g., rules); and individual (e.g., needs). These categories are explained below to illustrate the acquisition of expertise in a different knowledge paradigm from that of the original art form. The chapter thus exemplifies how creatives can develop multiple career pathways.

Environmental Constraints

Environmental constraints can be social, cultural or physical. For example, physical constraints could include gravity, temperature, light, humidity, and infrastructure such as building height or the type of floor on which a dance-teaching activity occurs. Social constraints could be family, mentors or peer groups. Cultural constraints could include contextual dance-related values, such as ethnological clothing used during performance (Kealiinohomoku, 1979), or specific dance genres, such as ballet or contemporary.

Mentors

A key social constraint identified by Rodrigues (2017) was mentors. Some participants described that their involvement with mentors, with a focus on pedagogy, influenced their effectiveness in becoming dance teachers. All mentoring relationships were observed to be informal, and not instigated by a third party (e.g., a person or an organisation) (Chao et al., 1992; Eby et al., 2007). According to Rodrigues (2017) the mentors' key role was to provide guidance, by exposing participants to significant teaching and learning experiences.

For example, a participant, Isabel, described how her mentors taught her to technically structure a ballet class using a philosophical approach to support the technical structure, which provided the basis for feedback as well:

> Melissa Stansfield and Steve Bacon...I worked with them...Melissa instructed me on how to structure a class. What goes into it and what comes next. How this happens. ...how I structure a class is very much based on their philosophies. ...they would watch and, then, they would give feedback on...what happened and why this exercise was probably less effective...all of those things.

Learning about her mentors' teaching *philosophies* might have exerted a significant influence on Isabel's pedagogical learning because her mentors would explore the rationale behind action. That is, why those events occurred: "what happened and why this exercise was probably less effective". More important is the fact that Isabel learned how to teach during real teaching events (e.g., "they would watch and, then, they would give feedback").

According to Warren (2006), the relationship between the individual and the environment allows the learner to explore information from the physical environment, in order to stabilise an intended behaviour. J.J. Gibson (1986) described this process as *education of attention*, the process representing a shift from non-specifying information to specifying information. This convergence, from non-specifying information to specifying information, as argued by Jacobs et al. (2001), happens after an individual's practice with feedback. This explanation suggests that the role of Isabel's mentors in her pedagogical development was to provide critical feedback to stabilise an intended behaviour.

Role models

Another social constraint identified by Rodrigues (2017) was role models. For instance, participants reported that their role models facilitated their perception of teaching behaviours, effective and ineffective teaching practices, and different ways for exploring teaching tasks. Participants described meaningful personal experiences in relation to "good" role models, whose practices to adopt, and "bad" role models, whose practices not to adopt.

In reference to "good" role models, a participant, Veronica, mentioned having started observing teachers during adulthood, while she was a professional performer. She described important features such as class structure; content; and teachers' approaches to learners in relation to expression, exploration and respect: "I think it is incredibly informative…what they are teaching you, how they structure a class, how they approach the class. …people who are clear; yet, open. …there is still some room for expression and experimentation. People who are respectful of learning, and learners."

Veronica's statements are in line with the role modelling literature that stated that a feature that distinguishes role models is the prevalence of specific attributes (e.g., role-expectation information, performance standards, and skill expertise) (Gibson, 2004). For example, Veronica might have perceived these role models as positive because they provided information concerning role-expectations (e.g., "how they structure a class, how they approach the class").

Furthermore, Veronica admired these teachers for respecting the "learning and the learners". These perceptions concerning role-expectation information and respect for learners suggest similarities concerning pedagogical goals between Veronica and those teachers, and have been reported within the role model literature by Lockwood and Kunda (1997), who suggested that individuals chose role models if they pursued the attainment of similar goals to those role models. Overall, role-expectation (Gibson, 2004) and goal similarity (Lockwood and Kunda, 1997) seem to have predisposed Veronica to choose a particular type of teacher as her role models: teachers who were simultaneously rigorous and flexible (role-expectation), and respectful of learners (goal similarity).

In reference to "bad" role models, Oscar, another participant in Rodrigues' (2017) study, reported that experiences with ineffective teachers also contributed to the acquisition of his pedagogical expertise because these teachers prompted him to move away from negative behaviours (Lockwood et al., 2002): "I have had that bad experience of having bad teachers, and bad role models. …teachers that do not really have a rapport with the students. …teachers that have been really bad communicators, badly organised…. So, I have used my 'bad' experience…to try and correct myself."

Experiencing these events seems to have caused a profound negative impact on Oscar; henceforth, Oscar evolved in the opposite direction of these "bad" role models. He prepared his classes thoroughly, and learned how to engage students and how to communicate effectively: "I always have my music prepared…have my classes prepared. I learnt how to engage the students, how to build their confidence. You do not just shout at them, you do not make them feel bad, you try to encourage them. …how to communicate…was a big focus."

Within ED, key concepts that explain behaviour are: attractors, which correspond to stable task solutions; repellers, which correspond to avoided states; and bifurcations, which correspond to behavioural transitions (Warren, 2006, p. 358). Oscar's perception of his previous teachers' behaviour forced him to change his behaviour. These changes correspond to behavioural states to be avoided and are categorised as repellers (Warren, 2006). Further, an individual needs to understand informational aspects that constrain behaviour because this information closes the acknowledged gap between perception and knowledge (Gibson, 1986).

This is exemplified by Oscar perceiving the lack of affect in his previous teachers' behaviour and judging them as negative role models. As a result, Oscar re-evaluated the experience, and his behaviour self-organised in the opposite direction. Oscar's example illustrates how the interaction of constraints – individual and environmental constraints (i.e., his intrinsic dynamics and the negative role model of a dance teacher) – allowed for knowledge creation and consequentially regulated his affect and pedagogical behaviour.

Students

Finally, the third environmental-social constraint reported by Rodrigues (2017) was students. Oscar, for instance, reported that his pedagogical development was influenced by teaching a range of students in specific contexts. He, in fact, made a conscious decision in choosing to teach children, adults and professional dancers. Each of these groups, according to Oscar, had different learning needs; consequently, he was required to develop specific strategies to address each group's learning needs. In terms of CLA, the diversity plus the contextual specificity of students acted as an interactional environmental constraint, altering Oscar's pedagogical effectiveness positively:

> I thought that, that [teaching children, adults and professionals] would give me more experience: if I know how to deal with children...with adults...with professional dancers because you have to change your approach to each group. So, I cannot teach, in the same way, children as I teach professional dancers. That teaches you to change your little techniques to target the audience in the best way.

Environmental constraints (i.e., students' variability) set boundaries on Oscar's pedagogical performance, emphasising that effective teaching is context dependent (Bullough and Baughman, 1995). This means that to be effective, Oscar needed to have acquired *knowledge of* (i.e., direct perception of environmental properties in relation to his body and action capabilities (Gibson, 1966)) and *knowledge about* (i.e., indirect perception through symbols, pictures and language (Gibson, 1966)) his students because it is the combination of both these types of knowledge that enables effective individual action (Araújo et al., 2009b).

Expertise, from an ED perspective, emphasises an individual's capability to be both stable and flexible during performance (Seifert et al., 2013). This further supports the functional role of interactional variability in determining which human coordinative structures emerge under the influence of constraints. Through exploration, a different range of behavioural organisational states are tested and the ineffective states are discarded (Davids et al., 2003; Turvey, 1990). Therefore, most probably, Oscar's interaction with the student variability, inherent in each group of learners, facilitated his understanding about each group having specific learning needs, which is why he revealed that each group needed to be pedagogically tailored (e.g., "I cannot teach, in the same way, children as I teach professional dancers").

Task Constraints: Rules

Task constraints relate to the goal of the task (Davids et al., 2008) and include rules (e.g., classical ballet aesthetic rules), instructions and equipment used (e.g., dance shoes). The information about the task that learners perceive is meaningful because it is typically used to support decision making, planning, and movement organisation (Davids et al., 2005). Therefore, task constraints have a strong influence on learners' intents, and are often the simplest factors to control or implement within a learning setting (Davids et al., 2005).

Dance teachers explore different dance genres (Gilbert, 2005), each genre possessing different task rules (Craine and Mackrell, 2010). For instance, ballet has codified movement and is generally considered to be more structured than contemporary (Craine and Mackrell, 2010). The different modes of structuring movement have different implications, for instance, pedagogical acquisition in ballet is attributed to persistence in preserving codified movement vocabularies (Birk, 2009; Morris, 2003), whereas, in contemporary, pedagogical acquisition is associated with exploring a variety of movement vocabularies (Foster, 1997).

Rodrigues (2017) suggested that the different rules involving ballet and contemporary influenced how dance teachers acquired pedagogical knowledge and skills. In reference to ballet, Hilda, for instance, reported that learning the technique and observing other ballet teachers' teaching was sufficient:

> Ballet is…a very formalised style. You can, sort of, learn technically, and have a good eye, and learn by watching. You can learn any teaching by watching… I am looking in the mirror and that is the ideal, and that is how I was taught. But, I would never teach that way, never.

Traditional processes of becoming a ballet teacher involve: acquiring ballet technique; perceiving previous teachers' teaching skills; and acquiring those skills through modelling (Lakes, 2005; Paskevska, 1992). However, the excerpt above suggests that although Hilda was taught to achieve the movement ideal through modelling, she did not embrace modelling in her teaching.

The affordances that an environment offers to an individual are conditional on the individual's skills and on her or his *form of life*, which refers to the stable and consistent patterns of behaviour of an individual in a certain context (Rietveld and Kiverstein, 2014). A form of life, within these terms, may be socio-culturally constrained because the skills that individuals acquire throughout participation in skilled practice are those skills needed to act in accordance with the norms (or rules, in the case of ballet) of that specific practice.

The way an individual develops skills through engagement with particular aspects of a particular environment, while being guided by more experienced individuals, refers to the normativity of affordances (Rietveld and Kiverstein, 2014). Hilda reported having learned how to teach from observing her previous teachers; for instance, about the use of imagery (e.g., "what Grace Morales had said, images that she had given…from imagery, and metaphor, and illusion"); effective choice of exercises (e.g., "exercises that I thought really made my body feel that it was being developed"); and how to sequence movement (e.g., "I learned how people sequenced movement"). This suggests that the perception of the pedagogical behaviour of Hilda's previous teachers was important to her awareness of which specific teaching features (i.e., affordances) could be important in order to become an effective ballet teacher.

Rietveld and Kiverstein (2014) further argued that the acquisition of skill is accomplished in socio-cultural environments and that, to some extent, there is agreement in what the members of a particular socio-cultural practice do. However, they additionally stated that the patterns of behaviour depend on an individual's continuous adjustments to the affordances of specific and concrete material settings (Rietveld and Kiverstein, 2014). Perhaps it was the need to continuously adjust to the affordances of specific and concrete material settings that prompted Hilda not to embrace modelling rules and the achievement of the movement ideal during teaching. For instance, Hilda reported that it was by experiencing principles of movement that she understood the need to explore other approaches for effective teaching: "teachers need to understand and underpin…these principles of movement…it was a way to think. How can I teach better, and how can I think about other approaches…".

The excerpt above suggests that Hilda's *adequacy of behaviour* (Rietveld and Kiverstein, 2014) was underpinned by aspects of embodied cognition, prompted by the perception and exploration of those "principles of movement". Furthermore, it would appear that Hilda's actions were goal-oriented: "…now you have to be able to do lots of things. So, your training [and teaching] needs to be adaptable. You need to train adaptive dancers. …if you look at the repertoire from ballet companies…everybody, all dancers, now, need this."

The excerpt above suggests that Hilda was able to perceive certain training and teaching needs emanating from ballet companies (e.g., "train adaptive dancers"); therefore, those "principles of movement" represented the way to achieve the goal of training adaptive dancers. This illustrates how Hilda's teaching practice was different from her previous ballet teacher's teaching practice, of aiming for the ideal movement form through modelling rules. Hilda's accounts suggest, therefore, that although her behaviour may resemble that of her previous teachers, acquired through socio-cultural practice, as Hilda matured, her individual form of life developed patterns of behaviour in relation to the affordances of her specific material settings.

In contrast to ballet, in contemporary, there is an endless possibility of seeking inspiration, for example, from emotions and concepts (Torrents et al., 2013b), which enhances new ways of exploring movement. On one hand, this freedom for exploring a range of movement inspirational possibilities provides the potential for the creation of new movement (Torrents et al., 2013a); nevertheless, on the other hand, it can cause instabilities in acquiring pedagogical content knowledge for teaching (Fortin et al., 2002). It would appear that the lack of rules in contemporary has created pedagogical instability and forced participants to search for teaching solutions. Rodrigues (2017) suggested that one way that his participants may have acquired pedagogical stable behaviours and expertise was by developing a kinaesthetic understanding.

For instance, Hilda reported that in the absence of rules to teach contemporary, she needed to improve her kinaesthetic understanding about the body and its functions.

> Because a lot of contemporary techniques are not standardised as ballet. ... Contemporary is...more a morphosis, more hybrid, people mix techniques a lot. ... there [are] so many things that you have to master. There are so many different kinds of ways where your centre of gravity, and all of that, changes. ...you need a very good underpinning of understanding the body and its function.

The excerpt above suggests that the non-standardised nature of contemporary allows for different combinations of different techniques; however, for an effective combination of techniques, a thorough understanding of the body and its functions is required. Fortin and colleagues (2002) argued that the movement of the body is validated by creating a real interaction between the self and the environment, so that the individual can learn from experiencing his or her own actions (Fortin et al., 2002). They argued that somatic techniques can be taught from a first-person viewpoint (i.e., without demonstration) or from a third-person viewpoint (i.e., with demonstration). The first-person viewpoint focuses on the student, and generates learning cultures that facilitate enquiry and organisation of experiences, according to individuals' unique ways of perceiving the environment. In contrast, the third-person viewpoint generates learning cultures that restrain individuality because this approach relies on repetition, replication, and on obedience to the teacher during teaching and learning events (Dragon, 2015). This implies that dance teachers' exploration of teaching tasks, from a first-person viewpoint, is more beneficial than from a third-person viewpoint for developing a kinaesthetic understanding, because the former facilitates the exploration of individual, task and environmental information that otherwise would not be perceived.

Individual Constraints: Needs

Individual constraints refer to the unique, physical, cognitive, physiological and emotional characteristics that can affect coordination and movement in learning and performance (Newell, 1986). For instance, weight, height, body morphology, genes, physical and technical abilities, as well as psychological characteristics such as cognition, needs, motivation and emotions, form the repertoire of individual constraints. It is important to understand individuals' needs because they relate to their motivation, intentions, goals and consequent behaviour to achieve those goals (Araújo et al., 2009a).

Rodrigues' (2017) study revealed that dance teachers' needs for pedagogical knowledge influenced their behaviour in searching for mentoring relationships. For instance, Veronica became aware that experienced dance teachers were associated with comprehensive pedagogical knowledge; thus, she searched for several individuals and established mentoring relationships.

> I have got a few mentors, different people for different reasons. ...some people are mentors not because they have taught me necessarily, but because of their experience in teaching dance: their scope of knowledge and understanding. So, it is more a quest. It is a mentor that I go to. ...I need some advice about this...

The excerpt above suggests that the rationale supporting Veronica's search for several mentors was to acquire pedagogical knowledge concerning specific issues (e.g., "I need advice about this") associated with each one's valid ways of teaching those issues (e.g., "because of their experience in teaching dance"). Lockwood et al. (2007) stated that the protégés' goals for mentoring relationships are usually aligned with the expertise that mentors need to possess. This appears to be in line with Veronica's rationale, above, for establishing mentoring relationships.

Self-Determination Theory (SDT) focuses on the understanding and explanation of intrinsic psychological processes that stimulate optimum health and functioning (Deci and Ryan, 2000). SDT investigates individuals' intrinsic developmental tendencies and inherent psychological needs, which generate the foundation for self-motivation and the integration of personality (Ryan and Deci, 2000). Deci and Ryan (2000) argued that to understand goal-directed behaviour and psychological development, the needs (e.g., for competence, relatedness and autonomy) that support goal achievement and that direct which regulatory processes guide individuals' goal pursuit, need to be addressed.

Within ED, an important feature influencing behaviour is goal-setting because goals influence control of action (Riccio and Stoffregen, 1988). Riccio and Stoffregen argued that goals impose important constraints on behaviour because they set the criteria for evaluating the interaction between the individual and the environment; furthermore, the authors showed that individuals adopt particular behaviours because those behaviours are key to goal achievement. SDT has also addressed the concept of goal-directed behaviour, separating the outcomes of goals, and "the regulatory processes through which the outcomes are pursued, making predictions for different contents and for different processes" (Deci and Ryan, 2000, p. 227).

To this extent, Veronica's need for pedagogical knowledge, within SDT, refers to an individual's innate needs for competence, relatedness and autonomy (Deci and Ryan, 2000). In Veronica's account, competence could be concerned with her desire to become an effective teacher; relatedness could be related to caring, and being cared for, by specific mentors; and, autonomy could perhaps be concerned with the self-organisation of her teaching behaviour in relation to her sense of self. According to SDT, Veronica's need for pedagogical knowledge would lead to her desire for obtaining pedagogical knowledge. To obtain this knowledge, she would set goals, which were reported in reference to her acquisition of specific knowledge from specific individuals (e.g., "I need some advice about this...different kinds of things float, resonate with me, from different people").

Goals of behaviour influence control of action because they determine the affordances of control strategies for perception and action (Riccio and Stoffregen, 1988), which are expressed through an individual's adaptive behaviour (Araújo et al., 2004). Therefore, an individual's adaptive behaviour can be understood as emergent, from the interaction of individual and environmental constraints, conditioned by particular goals (Araújo et al., 2004). As a consequence, Veronica's interactional goal-oriented behaviour would be to seek and engage in relationships with mentors (i.e., environmental-social constraints) and to seek specifying pedagogical information (i.e., affordances) useful to control pedagogical action related to her action capabilities.

Finally, intrinsic motivation refers to an individual's free choice in participating in interesting activities, without external influences, such as pressure, instructions, rewards or consequences (Deci, 1971); and are based on an individual's needs to feel competent and self-determined (Deci, 1975). This would explain that the satisfaction of Veronica's needs would be intrinsically motivated because she freely chose to be engaged in the mentoring processes. Veronica's account, therefore, exemplifies how CLA and SDT could be used to understand an individual's intrinsic motivated behaviour, imposed by the interaction of individual constraints (i.e., needs) and environmental-social constraints (i.e., mentors), in order to satisfy her needs of becoming pedagogically effective and self-determined through the development of goal-oriented activities.

CONCLUSION

This chapter describes individuals' behaviours and the acquisition of expertise, in situated environments, as emergent from the interaction of environmental, task and individual constraints (Davids et al., 2015; 2008; Seifert et al., 2013). Rodrigues' (2017) study highlights how environmental, task and individual constraints may have important implications for the acquisition of pedagogical expertise by dance teachers.

Within CLA, expert pedagogical performance may emerge throughout self-organisation processes when dance teachers interact with constraints. Interacting constraints allowed for dance teachers' perceptual attunement to critical information sources, which they used to regulate their actions in situated environments. Constraints are perceived as relevant sources of information, originating from the continuous interaction between the individual and the environment (Araújo et al., 2006); thus, adaptive goal-directed behaviour emerges when the individual tries to overcome constraints in order to reach a satisfactory solution (Chow et al., 2011). My argument illustrates how the CLA's multidisciplinary character can provide an understanding about how dance teachers' pedagogical expertise may emerge from situated environments, throughout the mediation of action and key constraints (Newell, 1986).

A number of broad implications for the acquisition of expertise, to enhance creative careers across a different knowledge paradigm (i.e., ED-CLA), can be formulated from Rodrigues' (2017) study. First, learning should include numerous opportunities for individuals to interact formally and informally with mentors and peers. Second, learning should include a range of different task constraints to allow natural evolution of expertise. Third, the design of learning experiences should explore *variability*, *by manipulating* a range of constraints; this way, learning environments would provide *opportunities to engage with the unexpected* and would facilitate discovery and the search for effective solutions.

The ability to perceive and become attuned to key affordances, in relation to an individual's *intrinsic dynamics*, appears to be critical for the acquisition of expertise because some environments may be more important than others in facilitating expertise development. Therefore, the fourth implication is to engage activities where individuals can explore their own *intrinsic dynamics and goal-setting*. Finally, above all, the design of expertise acquisition experiences should avoid the pursuit of a "one right answer" ideal expertise model.

NOTES

1. This chapter uses excerpts from Rodrigues (2017).
2. ORCID iD: 0000-0002-2382-6089, Faculty of Health, Queensland University of Technology, j.rodrigues@qut.edu.au.

REFERENCES

Araújo, D., Davids, K. and Hristovski, R. (2006). The ecological dynamics of decision making in sport. *Psychology of Sport and Exercise*, 7(6), 653–76.

Araújo, D., Davids, K., Bennett, S.J., Button, C. and Chapman, G. (2004). Emergence of sport skills under constraints. In A.M. Williams and N.J. Hodges (eds), *Skill acquisition in sport: Research, theory and practice* (pp. 409–43). London: Routledge.

Araújo, D., Davids, K.W., Chow, J.Y. and Passos, P. (2009a). The development of decision making skill in sport: An ecological dynamics perspective. In D. Araújo, H. Ripoll and M. Raab (eds), *Perspectives on cognition and action in sport* (pp. 157–69). New York, NY: Nova Science Publishers.

Araújo, D., Davids, K.W., Cordovil, R., Ribeiro, J. and Fernandes, O. (2009b). How does knowledge constrain sport performance? An ecological perspective. In D.

Araújo, H. Ripoll and M. Raab (eds), *Perspectives on cognition and action in sport* (pp. 119–131). New York, NY: Nova Science.

Baker, J., Wattie, N. and Schorer, J. (2015). Defining expertise: A taxonomy for researchers in skill acquisition and expertise. In J. Baker and D. Farrow (eds), *Routledge handbook of sport expertise* (pp. 145–55). Abingdon: Routledge.

Barab, S.A. and Kirshner, D. (2001). Guest editors' introduction: Rethinking methodology in the learning sciences. *The Journal of the Learning Sciences*, **10**(1–2), 5–15.

Berliner, D.C. (2001). Learning about and learning from expert teachers. *International Journal of Educational Research*, **35**(5), 463–82.

Birk, K. (2009). Pre-professional ballet training: Toward making it fit for human consumption. Paper presented at Global Perspectives on Dance Pedagogy: Research and Practice, Leicester, England.

Booth, E. (2003). Seeking definition: What is a teaching artist? *Teaching Artist Journal*, **1**(1), 5–12.

Bridgstock, R. (2015). KEY-shaped people, not T-shaped people: Disciplinary agility and 21st century work. Blog post. Retrieved from http://www.futurecapable.com/?p=102.

Bridgstock, R.S. and Goldsmith, B. (2016). Embedded creative workers and creative work in education. In R.S. Bridgstock, B. Goldsmith, J. Rodgers and G.N. Hearn (eds), *Creative graduate pathways within and beyond the creative industries* (pp. 37–55). New York, NY: Routledge (Taylor & Francis Group).

Bullough, R.V. and Baughman, K. (1995). Changing contexts and expertise in teaching: First-year teacher after seven years. *Teaching and Teacher Education*, **11**(5), 461–77.

Chao, G.T., Walz, P. and Gardner, P.D. (1992). Formal and informal mentorships: A comparison on mentoring functions and contrast with nonmentored counterparts. *Personnel Psychology*, **45**(3), 619–36.

Chow, J.Y., Davids, K., Hristovski, R., Araújo, D. and Passos, P. (2011). Nonlinear pedagogy: Learning design for self-organizing neurobiological systems. *New Ideas in Psychology*, **29**(2), 189–200.

Cooke, N.J. (1992). Modeling human expertise in expert systems. In R.R. Hoffman (ed.), *The psychology of expertise: Cognitive research and empirical AI* (pp. 29–60). Mahwah, NJ: Lawrence Erlbaum Associates.

Craine, D. and Mackrell, J. (2010). *The Oxford dictionary of dance* (2nd edn). New York, NY: Oxford University Press.Davids, K. and Araújo, D. (2010). The concept of "organismic asymmetry" in sport science. *Journal of Science and Medicine in Sport*, **13**(6), 633–40.

Davids, K., Araújo, D., Seifert, L. and Orth, D. (2015). Expert performance in sport: An ecological dynamics perspective. In J. Baker and D. Farrow (eds), *Routledge handbook of sport expertise* (pp. 130–44). Abingdon: Routledge.

Davids, K., Bennett, S.J. and Button, C. (2003). *Coordination and control of movement in sport: An ecological approach*. Champaign, IL: Human Kinetics.

Davids, K., Button, C. and Bennett, S. (1999). Modeling human motor systems in nonlinear dynamics: Intentionality and discrete movement behaviors. *Nonlinear Dynamics, Psychology, and Life Sciences*, **3**(1), 3–30.

Davids, K., Button, C. and Bennett, S. (2008). *Dynamics of skill acquisition: A constraints-led approach*. Champaign, IL: Human Kinetics.

Davids, K., Chow, J. and Shuttleworth, R. (2005). A constraints-based framework for nonlinear pedagogy in physical education. *Journal of Physical Education New Zealand*, **38**(1), 17–29.

Deci, E.L. (1971). Effects of externally mediated rewards on intrinsic motivation. *Journal of Personality and Social Psychology*, **18**(1), 105–15.

Deci, E.L. (1975). *Intrinsic motivation*. New York, NY: Plenum.

Deci, E.L. and Ryan, R.M. (2000). The "what" and "why" of goal pursuits: Human needs and the self-determination of behavior. *Psychological Inquiry*, **11**(4), 227–68.

Dragon, D.A. (2015). Creating cultures of teaching and learning: Conveying dance and somatic education pedagogy. *Journal of Dance Education*, **15**(1), 25–32.

Eby, L.T., Rhodes, J.E. and Allen, T.D. (2007). Definition and evolution of mentoring. In T.D. Allen and L.T. Eby (eds), *The Blackwell handbook of mentoring: A multiple perspectives approach* (pp. 7–20). Malden, MA: Blackwell Publishing.

Fortin, S., Long, W. and Lord, M. (2002). Three voices: Researching how somatic education informs contemporary dance technique classes. *Research in Dance Education*, **3**(2), 155–79.

Foster, S.L. (1997). Dancing bodies. In J.C. Desmond (ed.), *Meaning in motion: New cultural studies of dance* (pp. 235–58). Durham, NC: Duke University Press.

Gibson, D.E. (2004). Role models in career development: New directions for theory and research. *Journal of Vocational Behavior*, **65**(1), 134–56.

Gibson, J.J. (1966). *The senses considered as perceptual systems*. Boston, MA: Houghton Mifflin.

Gibson, J.J. (1986). *The ecological approach to visual perception*. Hillsdale, NJ: Lawrence Erlbaum Associates.

Gilbert, A.G. (2005). Dance education in the 21st century: A global perspective. *Journal of Physical Education Recreation and Dance*, **76**(5), 26–35.

Hearn, G. and McCutcheon, M. (2020). The creative economy: the rise and risks of intangible capital and the future of creative work. In G. Hearn (Ed.), *The future of creative work: Creativity and digital disruption* (pp. 14–33). Cheltenham, UK and Northampton, MA, USA: Edward Elgar Publishing.

Huddy, A. and Stevens, K. (2011). The teaching artist: A model for university dance teacher training. *Research in Dance Education*, **12**(2), 157–71.

Jacobs, D.M., Runeson, S. and Michaels, C.F. (2001). Learning to visually perceive the relative mass of colliding balls in globally and locally constrained task ecologies. *Journal of Experimental Psychology: Human Perception and Performance*, **27**(5), 10–19.

Kauffman, S.A. (1993). *The origins of order: Self organization and selection in evolution*. New York, NY: Oxford University Press.

Kealiinohomoku, J. (1979). You dance what you wear, and you wear your cultural values. In J.M. Cordwell and R.A. Schwarz (eds), *The fabrics of culture: The anthropology of clothing and adornment* (pp. 77–83). The Hague: Mouton.

Kelso, J.A.S. (1991). Anticipatory dynamical systems, intrinsic pattern dynamics and skill learning: Reaction to Bullock and Grossberg, 1991. *Human Movement Science*, **10**, 93–111.

Lakes, R. (2005). The messages behind the methods: The authoritarian pedagogical legacy in Western concert dance technique training and rehearsals. *Arts Education Policy Review*, **106**(5), 3–20.

Lockwood, A.L., Evans, S.C. and Eby, L.T. (2007). Reflections on the benefits of mentoring. In T.D. Allen and L.T. Eby (eds), *The Blackwell handbook of mentoring: A multiple perspectives approach* (pp. 233–6). Malden, MA: Blackwell Publishing.

Lockwood, P. and Kunda, Z. (1997). Superstars and me: Predicting the impact of role models on the self. *Journal of Personality and Social Psychology*, **73**(1), 91–103.

Lockwood, P., Jordan, C.H. and Kunda, Z. (2002). Motivation by positive or negative role models: Regulatory focus determines who will best inspire us. *Journal of Personality and Social Psychology*, **83**(4), 854–64.

Morris, G. (2003). Problems with ballet: Steps, style and training. *Research in Dance Education*, **4**(1), 17–30.

Newell, K.M. (1986). Constraints on the development of coordination. In M.G. Wade and H.T.A. Whiting (eds), *Motor development in children: Aspects of coordination and control* (Vol. 34, pp. 341–60). Dordrecht: Martinus Nijhoff.

Paskevska, A. (1992). *Both sides of the mirror*. New York, NY: Princeton Book Company.

Patton, M.Q. (2002). *Qualitative research and evaluation methods*. Thousand Oaks, CA: Sage.

Phillips, E., Davids, K., Renshaw, I. and Portus, M. (2010). Expert performance in sport and the dynamics of talent development. *Sports Medicine*, **40**(4), 271–83.

Riccio, G.E. and Stoffregen, T.A. (1988). Affordances as constraints on the control of stance. *Human Movement Science*, **7**, 265–300.

Rietveld, E. and Kiverstein, J. (2014). A rich landscape of affordances. *Ecological Psychology*, **26**(4), 325–52.

Rodrigues, J. (2017). *The acquisition of pedagogical expertise in dance: A constraints-led approach* (Unpublished doctoral dissertation). Queensland University of Technology, Australia, Brisbane.

Ryan, R.M. and Deci, E.L. (2000). Self-determination theory and the facilitation of intrinsic motivation, social development, and well-being. *American Psychologist*, **55**(1), 68–78.

Seifert, L., Button, C. and Davids, K. (2013). Key properties of expert movement systems in sport. *Sports Medicine*, **43**(3), 167–78.

Shulman, L.S. and Shulman, J.H. (2004). How and what teachers learn: A shifting perspective. *Journal of Curriculum Studies*, **36**(2), 257–71.

Swann, C., Moran, A. and Piggott, D. (2015). Defining elite athletes: Issues in the study of expert performance in sport psychology. *Psychology of Sport and Exercise*, **16**, 3–14.

Torrents, C., Castañer, M., Dinušová, M. and Anguera, M.T. (2013a). Dance divergently in physical education: Teaching using open-ended questions, metaphors, and models. *Research in Dance Education*, **14**(2), 104–19.

Torrents, C., Castañer, M., Jofre, T., Morey, G. and Reverter, F. (2013b). Kinematic parameters that influence the aesthetic perception of beauty in contemporary dance. *Perception*, **42**(4), 447–58.

Turvey, M.T. (1990). Coordination. *American Psychologist*, **45**(8), 938–53.

Warren, W.H. (2006). The dynamics of perception and action. *Psychological Review*, **113**(2), 358–89.

Withagen, R., de Poel, H.J., Araújo, D. and Pepping, G.J. (2012). Affordances can invite behavior: Reconsidering the relationship between affordances and agency. *New Ideas in Psychology*, **30**(2), 250–58.

15. Do creative skills future-proof your job? Creativity and the future of work in an age of exponential technological advancement[1]

Ruth Bridgstock,[2] Russell Tytler[3] and Peta White[4]

INTRODUCTION

It is the year 2030. Jacob has just embarked on his third career change in 15 years, and is enrolled in a postgraduate course with the aim of becoming a nostalgist, which is one of a "new wave" of creative job roles that came about with massive advances in digital technologies. People now live longer and there is a strong demand in virtual reality experiences that allow the elderly to relive the best parts of their lives. Nostalgists combine interior design, digital research and virtual-reality development to create these experiences (Tytler et al., 2019). While Jacob is studying, he is also working part-time as a sub-contracted aged-care worker, for which he was awarded a foundational qualification last year. Prior to this, Jacob worked as a plasterer, but it was becoming too hard to compete with automated plastering services, and so he decided to change.

Jacob's story was created from themes in research interviews conducted in 2019 with experts in science, technology and social trends. While specifics such as job titles may not stand the test of time, the presence of consistent themes in the interviews helps to clarify how work and careers might unfold in the future, and which capabilities are likely to be important. This chapter shares the findings of those interviews.

The speed, extent and impact of technological changes on the world of work have been subjected to much debate during the last few years. Popular media, government and consultancy reports, and academic literature alike abound with deterministic predictions of "robots coming for our jobs", arguing that between 20 per cent and 60 per cent of job roles will disappear over coming decades under the influence of automation, robotics and artificial intelligence (AI). Others emphasise the creation of entirely new jobs that are based on technological advances. A third category of future-focused literature emphasises qualitative and task-based changes to job roles. As machines become able to perform some tasks better and more cost-effectively than humans, human jobs are predicted to become either more repetitive and low-skilled, or more complex and high-skilled. Work arrangements are already changing, with movement from a "traditional" model of full-time, continuing employment activity at a place of work to employment arrangements that are more flexible in a number of ways.

Creativity is one category of capability that has been suggested to be uniquely human, or at least less "susceptible" to automation than other categories. In turn, job roles that involve significant creativity may also be less susceptible to automation. According to some sources, and in line with contemporary arguments about the creative economy (Bakhshi et al., 2015; Cunningham and Potts, 2015), such job roles may continue to add greater value in the future world of work than other roles. This chapter considers the proposition that creative skills and tasks "future-proof" human work roles – or put another way, that creative workers possess distinctive and valuable creative capabilities that are unlikely to be replaced by machines. The discussion includes the likelihood that machines can become, or are becoming, creative in different ways and contexts, as well as exploring the ways that humans are creative, and how creativity adds value to our economy and society, now and into the future.

This chapter commences by exploring the contrasting predictions about technological advances and the future of work, along with the economic and social mediating forces that are thought to be influential. Then, it discusses different kinds of creativity and other "uniquely human" capabilities in terms of the types of value that they add, and draws upon the study *100 jobs of the future* to characterise the potential importance and value of creativity and other skills in a highly technology-dominated world of work. Finally, the chapter asks some larger questions about the complementary roles that creative humans and machines might be able to play in meaningful future activities, such as critical and creative reasoning and design, which promote economic and environmental sustainability, and strengthen individual and community health and well-being (Gershenfeld et al., 2017).

TECHNOLOGICAL ADVANCES AND THE CHANGING FUTURE (AND PRESENT) OF WORK

Scholars argue that we are now experiencing the fourth industrial revolution (Schwab, 2016), also known as the second machine age (Brynjolfsson and

McAfee, 2014) and the third digital revolution (Gershenfeld et al., 2017). Such scholars suggest that we stand on the brink of an era in which AI, automation, robots and digital networks will transform the way we live, work and learn. Changing work roles and activities due to technological progress is not a new phenomenon. The Luddites of the first industrial revolution, afraid for their jobs, destroyed weaving machinery. In the 1970s and 1980s, thousands of typesetting compositors (workers who arranged movable type for printing presses) were made obsolete by computerisation. The difference in the early twenty-first century seems to be that a diverse range of technological advancements are occurring both simultaneously and at an exponential rate, creating much larger, and more widespread, changes to work than ever before.

Disappearing Jobs

Over the last decade, discussions of work and employment have become dominated by predictions about technology-driven disappearance or creation of jobs, changes to tasks inside jobs, and change in the organisation and experiences of future work. Frey and Osborne's (2013) working paper set the tone for many of the initial predictive studies, suggesting that, based on the characteristics of existing roles and the predicted capabilities of computers over coming years, nearly half of the jobs in the United States (US) were at high risk of being automated in the next 20 years. This study was followed by several others using similar methodologies demonstrating that the technical potential for automation of job roles in the next few years is quite high. For instance, in Australia, in the next decade or so, around 40 per cent of jobs are forecast to be strongly affected by computerisation and automation (Durrant-Whyte et al., 2015; Edmonds and Bradley, 2015). Looking at current work tasks rather than roles, Manyika et al. (2017) indicated that about half of work activities could be automated by 2030.

Rearranging Jobs

Literature also suggests that digital technologies will continue to change how work is arranged and experienced by workers. Retail, service and entertainment functions have already started to shift online or be offered through distributed networks. The gig economy relies on digital technologies to mediate between companies and workers who engage in freelance and casual work, and the ironically named "sharing economy" enables commercial transactions between workers and customers, through platforms such as Airtasker, Uber and Airbnb. Crowdsourcing work platforms such as Amazon's MTurk distribute labour-intensive work to many unrelated workers.

These new work arrangements have been praised for their potential to enhance work–life balance and to offer flexibility to workers in terms of supporting them to set their own hours and conditions. However, such arrangements have also been strongly criticised for compromising job security and not offering leave and other benefits, thus, shifting many of the risks associated with employment onto the worker (World Economic Forum, 2018). Even in more "conventional" employment situations where organisations employ workers directly, there may be a shift towards outsourcing of work, short and flexible contracts, and freelancing – all directly or indirectly enabled by digital technologies.

Emerging and Newly Appearing Jobs

While many frame the future of work pessimistically, others have pointed out that technology can enhance work lives, eliminating dangerous or boring work, and creating new job roles (see Lent, 2018). Citing the example of the iPhone, Huws (2014) discusses how new technologies can create and support innumerable work roles right across the value chain and across many countries, from raw materials through to design, manufacture, sale and disposal. The same kinds of online distributed networks that are rearranging jobs are diversifying options for consumers and businesses to access goods and services, offering new avenues for commodification, and more jobs. Autor (2015) discusses the complementarities between automation and human labour that can actually increase productivity, earnings, and demand for workers.

In a striking counter-balance to the "robots coming for our jobs" narrative, a sub-set of the work futures literature is now devoted to how new technologies have the potential to create new and exciting job roles that have not yet been thought of (e.g., Borland and Coelli, 2017). This literature cites examples of job roles and entire fields where this has already occurred in the last few years, such as visual effects (VFX) compositors[5] (in contrast to the printing press compositors already mentioned), other VFX artists, social media marketers and cybersecurity professionals. Tytler et al. (2019) recently published a study titled *100 jobs of the future*, which aimed to inform educational curriculum development by interrogating technological and research advances, along with social and economic trends, to predict and describe potential job roles for the next 30 years. The study then analysed the key capability requirements for the predicted work roles. Tytler et al.'s (2019) study will be discussed in more detail later in this chapter, in relation to creativity and creative capabilities for future work.

Influences Beyond Technical Potential: Jobs Disappearing, Appearing and Rearranging

There is a significant lack of accord among scholars about the impact of technological change on the future of work in the twenty-first century. In part, this discord is because of the very complex influences that social and economic forces have on work, and also on technology development and application. For instance, Frey and Osborne (2013) moderated their initial estimate of a nearly 50 per cent role disappearance by documenting a range of forces that could slow or increase the pace of disruption of jobs by technology, such as regulation and policy, economic conditions and public opposition. Similarly, Spencer (2018) considers the "politics of production" that shape how technology operates in the economy, emphasising the role of ownership and the current political economy of capitalism on how technological advances and the future of work evolve. For instance, to the extent that people might be displaced by machines and, thus, be unemployed or underemployed with little or no income, they will also consume less. Beyond technology per se, such factors are potentially capable of modifying the primary drivers of large-scale social and economic change.

Mishel and Bivens (2017) discuss the importance of economic and social influences on work rather than an inexorable future "robot apocalypse" (p. 3). Specifically, the authors analyse the current state of income polarisation in the US and show that, to date, automation has actually had very little to do with this. Mishel and Bivens (2017) argue that rapid technological advances historically have been associated with some displacement of workers, but better outcomes for them overall. The authors do not discount the probability of technologically induced shifts in work and jobs in the future, but indicate that rapid technological advances need to be complemented by labour policies to improve equity and labour standards for workers.

Other discussions of the power of social and economic influences on the future of technology and work explore the rise in offshoring, migration and global employment practices, along with exploitation of tax opportunities and exposure (Ford, 2015; Neufeind et al., 2018). Such trends could lead to a smaller and more precarious workforce; however, at the same time labour shortages, for example, through the retirement of baby boomers, could compensate (e.g., Lent, 2018).

All of the Above (and Maybe More)? A Complex Picture of the Future of Work

Scholars tend to argue strongly for one position or another, but both "optimistic" and "pessimistic" views of technological advances and the future of work could well come to pass (e.g., Lent, 2018), possibly concurrently, while continuing to be mediated strongly by social and economic influences. Routine tasks are vulnerable to automation – and, indeed, some of these jobs (such as straightforward customer service and manufacturing roles) are already in decline (Autor et al., 2003; Frey and Osborne, 2013). However, some authors (Spencer, 2018; Spreitzer et al., 2017) propose that some jobs in the future, at least for a while, may be even more routine and repetitive. These are job roles that are very straightforward for humans, but are beyond the capabilities of machines, or for which it may be cheaper and quicker to employ a human being. For instance, some of the tasks that MTurk employs humans to do, often for a few cents each, involve categorising data, captioning pictures, tagging metadata, analysing sentiments and placing ads in videos. Many of these tasks then feed machine-learning algorithms and teach AI systems. Workers who engage in these tasks, often through crowdsourcing platforms, are predicted to make up a large proportion of the human "precariat" in the future (Standing, 2011).

While some scholars have assumed that today's white-collar or knowledge workers are likely to occupy highly valued roles in the future, studies have shown that these workers are not immune to automation. Susskind and Susskind (2016) explore the future of the professions, and demonstrate that a significant part of many of these roles can be automated. As Hughes (2017) points out, nurses with computers may provide cheaper diagnoses than doctors. On the other hand, there may also be created some more "fulfilling jobs" for humans that require a wider range of more complex skills to either work with, complement or circumvent technology. The nurse who trains to work with computer diagnosis may be one case in point.

In all these analyses, there seem to be a set of "uniquely human" skills, which, at least in the next few years, will continue to be highly valued while also being less susceptible to automation. Like the workers who will undertake the routine and repetitive tasks, these professional workers may often be doing what is beyond the capabilities of machines or will complement machine activities (Spreitzer et al., 2017). However, unlike the workers who will undertake the routine and repetitive tasks, these professionals' work and capabilities will be highly valued and will be associated with premium wages and working conditions, and a high level of agency in shaping their work and careers (Spreitzer et al., 2017).

"UNIQUELY HUMAN" SKILLS?

As technological advances continue, automation will no longer be limited to carrying out tasks that are repetitive and simple. Sophisticated machine-learning systems are starting to be able to accomplish high-level tasks that until recently could be accomplished only by humans, or could not be accomplished at all (Schwab, 2016). Frey and Osborne (2013) suggested that pattern-recognition processes (involving machines learning to classify data in sophisticated ways) would be central to allowing machines to compete with humans in many tasks.

Frey and Osborne (2013) did envision a range of non-routine work activities that were much less likely to be replicated by machines over the next few years, including those involving social and emotional intelligence (e.g., negotiation, care, and persuasion of humans), complex perceptual and manual tasks, and creativity. Autor (2015) and Edmonds and Bradley (2015) also list roles that involve complex problem-solving, interpersonal skills and "creative intelligence" as being less automatable in the short-to-medium term.

The notion of highly valued human skills is supported by a number of studies of changes in the skill composition of work roles over the last 15 years. MacCrory et al. (2014) examined how roles had shifted in the US between 2006 and 2014. The authors found that, in line with Frey and Osborne (2013), there has been an increase in demand for interpersonal skills and skills required to use and manage machines. In 2015, the World Economic Forum (2016) surveyed employers, and found a predicted sharp increase in demand by the year 2020 for high-level cognitive capabilities, systems skills, and complex problem-solving skills such as mathematics, logical reasoning, visualisation, systems analysis and creativity.

CREATIVITY: SAFE OR NOT?

For many scholars, creativity is one of the defining characteristics of what it means to be human, and it tends to be one of the less controversial items in the lists of capabilities suggested to be relatively "safe" from automation. That said, there continues to be conflict over exactly what creativity means, and which types of creativity might be more "safe". As Eliza Easton (2018) from the United Kingdom (UK) innovation foundation the National Endowment for Science, Technology and the Arts (Nesta) argues, in many quarters, the term has become so overused as to be meaningless. Easton cites Subway's servers, known as "sandwich artists", as an example of this.

Recent Nesta studies engage explicitly with the question of whether human creative jobs are machine-proof. In *Creativity vs. robots* (Bakhshi et al., 2015), the authors used a methodology similar to that of Frey and Osborne's (2013) study to predict the likelihood of automation of creative job roles. Bakhshi et al. (2015) defined creativity quite broadly, as "the use of imagination or original ideas to create something" (p. 5). Each job role was coded in terms of the probability that it was creative. High-probability creative jobs included creative industries roles that are based on the production of cultural/creative content or services (such as creative artists, filmmakers and architects). The study also included several management occupations, including marketing and sales directors, and advertising account managers. Also included as high-probability creative roles were a range of computer, engineering and science occupations, such as civil engineers, IT specialist managers and chemical scientists. In the US, 21 per cent of all roles were calculated to be high-probability creative roles, and in the UK, 24 per cent were high-probability creative roles. In both countries, automation posed minimal risk to 86 per cent of these workers.

In *Creativity and the future of skills* (Easton and Djumalieva, 2018), researchers mapped where mention of creativity and creative skills was made in 35 million job advertisements in the UK. The study then used previous research that relied on a combination of machine learning and expert judgement to predict which jobs were most and least likely to grow by 2030 (Bakhshi et al., 2017) to show that jobs asking for creativity were far more likely to grow as a percentage of the labour market than other job roles. Other transferable skills, such as detail orientation, basic computer skills and customer service skills, were negatively correlated with predicted job growth. By comparison, creative skills paired with digital skills (for non-routine tasks, problem-solving and creating digital content), along with creative skills paired with organisational skills, research and project-management skills, were powerful predictors of job role growth.

Predicting Future Jobs and Skills: 100 Jobs of the Future

The Nesta studies are useful in starting to unpack the predicted value of different types of human creativity in the context of a future world of work dominated by technological advancement. A very recent study conducted by Tytler et al. (2019) used a different methodology to explore the kinds of job roles that may exist in 20 years (either newly created or pre-existing). Thematic analysis of the capabilities needed for these job roles reveals more detail about the extent to which, and ways in which, human creativity might be valued in the future.

The researchers recruited experts across technological and research advancements, and domains of application and industries critical to future work. These areas included health, agriculture, engineering and materials science, transport and mobility, computing and AI, commerce, and education. The experts were interviewed, exploring trends in their areas of expertise, potential future job roles, tasks associated with these roles, and the skills and capabilities that young people might develop to prepare for such roles. Transcripts of the interviews were analysed to identify exemplar job roles that captured key elements of what the experts were describing. Overall, 100 job descriptions were constructed that extracted the key trends and advancements, roles and tasks, and required capabilities represented in the expert interviews.

The project does not make quantitative predictions about the likelihood or extent of future job roles, nor does it attempt to be exhaustive in its predictions. Its findings can be useful in describing qualitatively the kinds of human creativity that may be important in the future, along with the relationships between creativity and technology, and with other capabilities. The findings are summarised in Table 15.1.

Table 15.1   100 jobs of the future: jobs coded by themes

Theme	
Creative job	82
Primarily creative industries job	21
Primarily STEM job	58
Transdisciplinary job	65
Technology advancement creative job	15
Technology input creative job	57
Creative job with technology distribution	72
Creative job with both technology input and distribution	35
Creative job with no technology elements	1
Creative job involving collaboration	74
Creative job involving social intelligence	82
Creative job involving organisational skills / project management	82

Note: $N = 100$. Please note that these categories overlap and do not therefore add to 100 jobs.
Source: Tytler et al. (2019).

In total, 82 of the job roles were hand-coded as high-probability creative, using the Bakhshi et al. (2015) definition relating to the use of imagination or original ideas to create something, with the addition that the created thing or experience be useful or valued in some way. Of the 82 high-probability creative jobs, one-quarter (21 jobs) fell within the current definition of creative industries roles (Department for Culture, Media and Sport, 2015), focusing on production of cultural/creative content or services. Creative industries roles included various kinds of designers (autonomous vehicles, 3D (three-dimensional) printed buildings, gamification, augmented reality experiences), marketers (personalised marketing), content creators (personal brand content) and artists (e.g., swarm artist, who uses swarms of hundreds of drones moving in formation to create art, music or performance-based cultural experiences). A total of 58 of the high-probability creative jobs tended to fall within STEM (science, technology, engineering and mathematics) categories, broadly defined. STEM roles included engineers (bioprinting, mechatronics, additive manufacturing, nanomedical, weather), scientists (terraforming microbiology, biofilm, entomicrobiotech, water management), data or data systems roles (automation anomaly analysis, forensic data analysis, data privacy strategy) and sophisticated kinds of programmers (quantum computers). While some job descriptions in the STEM category called explicitly for "creativity" or "creative skills", they were more likely to call for "problem-solving skills" to create solutions, things or experiences.

Most high-probability creative roles would be categorised primarily into creative industries and STEM roles, but some roles defied disciplinary categorisation. For instance, the "fusionist" role was described as someone who "designs approaches to bring together professionals from art, engineering, research, science, and other disciplines to create innovative ideas, experiences, and solutions to complex problems. Fusionists will be employed across many industries, where they will act as bridges between people with specialist disciplinary knowledge." (Tytler et al., 2019, p. 58).

"Trendwatchers" were described as:

> those who will know what is likely to happen next, and how to make the most of it. In a future where the pace of change is incredibly rapid, they will be employed by big companies and government agencies to watch the latest developments in science, technology, social issues, and the environment. Trendwatchers will be across the future developments in multiple areas, as new opportunities often occur at the intersection of trends (for example, the development of DNA-based data storage technologies that increase data storage capacity exponentially, at the same time as swarm technologies that allow very big data approaches to city surveillance). (Tytler et al., 2019, p. 60)

Even the high-probability creative roles that could be categorised primarily into creative industries or STEM tended to have strong transdisciplinary elements, with 65 of the 82 jobs requiring capabilities across different disciplinary areas. One example of this is the digital implant designer, who creates "body hacks" that will be implanted into people's bodies and brains to ensure their health and enhance their lifestyles. Digital implant designers therefore have knowledge of design processes, technical skills relating to implant technologies and digital systems integration, and knowledge of human biology and medical advances.

The high-probability creative roles were also coded in terms of whether they: (a) created technological advances; (b) used technological advances as inputs into creative activities or to distribute creative outputs; or (c) did not rely on technological advances at all – the creative elements of their work were entirely analogue. A total of 15 of the 82 high-probability creative roles created technological advances, through engineering, scientific discovery and/ or design processes. Another 57 drew upon technological advances as inputs into creative activities, with a total of 72 relying on technology to distribute creative outputs. Thirty-five roles used technological advances as both inputs and for distribution. Only one creative job role did not rely on technological advances at all: the "analogue experience guide", who creates experiences for people to "unplug from digital life and reconnect with the natural world, without digital implants or augmented reality. Analogue Experience Guides help people to appreciate a simpler and slower life by experiencing natural environments such as forests or mountainous areas without digital infrastructure or surveillance" (Tytler et al., 2019, p. 91).

Finally, the high-probability creative roles were coded to identify related transferable skills. In line with suggestions by Frey and Osborne (2013) and Bakhshi et al. (2015), social intelligence in different forms was important in the vast majority of high-probability creative jobs in this study. A total of 74 of the creative jobs required collaboration with others, and the vast majority of these collaborative relationships involved inputs from people in different roles and from different disciplines. For instance, "multisensory experience designers":

> bring together virtual reality, haptic and biofeedback / biometric technologies to create fully immersive basis of games and leisure activities, marketing campaigns, and education / training. They work in teams with specialists in each of these fields to create experiences that are indistinguishable from the physical world, or alternatively that are incredibly different from the physical world, but are still "hyperreal". (Tytler et al., 2019, p. 77)

As in the Bakhshi et al. (2015) study, the 74 creative jobs suited project teams that Caves (2000) described as "motley crews", with diverse inputs, a combination of creative and straightforward tasks, and complicated modes of organisation. Thought about like this, it can be seen how creative project-based collaborations (in creative industries, STEM, or transdisciplinary teams) might be resistant to automation. Further, all 82 creative roles (and a total of 97 roles across the whole data set) required social intelligence of some kind, in terms of interacting with other humans beyond any collaborative working arrangements. The social intelligence required differs by role, but involves understanding the needs of others in order to solve problems (nostalgist, cross-cultural capability facilitator), to persuade (negotiator for rights to intellectual property of AI), to sell or market (aesthetician, data commodities broker), to teach (lifelong education adviser) and/or to care and support (100-year counsellor). All 82 creative roles (and a total of 96 across the whole data set) also required organisational and project-management capabilities of some kind, echoing the findings in the Easton and Djumalieva (2018) study.

CREATIVE SKILLS AND CREATING THE FUTURE

This chapter has explored some of the deeply divided scholarly opinions about the impact of technological advancements on the future of human work. While some scholars have emphasised the disappearance of job roles under the influence of automation, AI and robotics (some of which has already commenced), others emphasise the creation of new roles that work in concert with new technologies. There is somewhat more agreement about future changes to work arrangements and experiences of work and careers, and also the fact that potential disruptions to work are strongly mediated by the social and economic context. Several authors now suggest that we should not be worried about a relentless "robot apocalypse", but rather the possibility, under capitalist systems, that the opportunities afforded by technology will be exploited to concentrate wealth, increase income inequality and perpetuate precarious labour conditions.

There do seem to be some capabilities that humans possess that machines do not – at least in the short-to-medium term – and these may be valued in terms of job roles in the future. Creativity is one of these capabilities (although it may be possible in the reasonably short term for machines to be valued for their creativity also, which is probably distinct from ours). The human creativity in question is a "generic" type of creativity, that is, the ability of people to create something new that is valued. Creativity of this type can be expressed in the context of creative industries or STEM skills, and often will rely on transdisciplinarity, collaborative processes, social intelligence, and high levels of organisation and project management. This creativity works using technological advances in various ways – through creation of new technologies, technological inputs or technological means of distribution.

In thinking about the challenges that we face in the coming decades, it may be worthwhile to pivot away from the question of whether jobs may disappear or appear. Rather, it may be useful for us all to think more broadly about how creative humans can work together with machines to promote environmental and economic sustainability, and the well-being of people and communities (Gershenfeld et al., 2017). As Riel Miller (2018), Chief Futurist at UNESCO suggests, rather than trying to forecast or react to a potential future, it is possible to create preferred futures for humans and machines through our considered choices and actions.

NOTES

1.   This chapter includes excerpts from Tytler et al. (2019). The funding of Deakin University, Griffith University and Ford Australia is gratefully acknowledged.

2. ORCID iD: 0000-0003-0072-2815, Centre for Learning Futures, Griffith University, r.bridgstock@griffith.edu.au.
3. ORCID iD: 0000-0003-0161-7240.
4. ORCID iD: 0000-0002-0225-5934.
5. Good VFX compositors are very much in demand in the twenty-first century world of work. These are the VFX artists who integrate live-action and digital elements to create the final completed shot in a film or game.

REFERENCES

Autor, D. (2015). Why are there still so many jobs? The history and future of workplace automation. *Journal of Economic Perspectives*, **29**(3), 3–30.

Autor, D., Levy, F. and Murnane, R.J. (2003). The skill content of recent technological change: An empirical exploration. *Quarterly Journal of Economics*, **118**, 1279–333.

Bakhshi, H., Downing, J.M., Osborne, M. and Schneider, P. (2017). *The future of skills: Employment in 2030*. London: Pearson.

Bakhshi, H., Frey, F. and Osborne, F. (2015). *Creativity vs. robots*. Nesta, April. Retrieved from Nesta website: https://media.nesta.org.uk/documents/creativity_vs._robots_wv.pdf.

Borland, J. and Coelli, M. (2017). Are robots taking our jobs? *Australian Economic Review*, **50**(4), 377–97.

Brynjolfsson, E. and McAfee, A. (2014). *The second machine age: Work, progress, and prosperity in a time of brilliant technologies*. New York, NY: Norton.

Caves, R. (2000). *Creative industries: Contracts between art and commerce*. Cambridge, MA: Harvard University Press.

Cunningham, S. and Potts, J. (2015). Creative industries and the wider economy. In *The Oxford Handbook of Creative Industries* (pp. 387–404). Oxford: Oxford University Press.

Department for Culture, Media and Sport (2015). Creative industries economic estimates January 2015 statistical release. Retrieved from https://assets.publishing.service.gov.uk/government/uploads/system/uploads/attachment_data/file/394668/Creative_Industries_Economic_Estimates_-_January_2015.pdf.

Durrant-Whyte, H., McCalman, L., O'Callaghan, S., Reid, A. and Steinberg, D. (2015). The impact of computerisation and automation on future employment. In *Australia's Future Workforce?* (pp. 56–64). Melbourne: Committee for Economic Development of Australia.

Easton, E. (2018). Is creativity the key to the job market of the future? British Council. Retrieved from https://www.britishcouncil.org/anyone-anywhere/explore/digital-creativity/job-market-future.

Easton, E. and Djumalieva, J. (2018). *Creativity and the future of skills*. Nesta. Retrieved from Nesta website: https://media.nesta.org.uk/documents/Creativity_and_the_Future_of_Skills_v6.pdf.

Edmonds, D. and Bradley, T. (2015). Mechanical boon: Will automation advance Australia? (Office of the Chief Economist Research Paper No. 7/2015). Canberra: Australian Government Department of Industry, Innovation and Science.

Ford, M. (2015). *Rise of the robots: Technology and the threat of a jobless future*. New York, NY: Basic Books.

Frey, C.B. and Osborne, M.A. (2013). The future of employment: How susceptible are jobs to computerisation? Working paper, University of Oxford. Retrieved from http://sep4u.gr/wp-content/uploads/The_Future_of_Employment_ox_2013.pdf.

Gershenfeld, N., Gershenfeld, A. and Cutcher-Gershenfeld, J. (2017). *Designing reality: How to survive and thrive in the third digital revolution.* New York, NY: Basic Books.

Hughes, J. (2017). What is the job creation potential of new technologies? In K. LaGrandeur and J. Hughes (eds), *Surviving the machine age* (pp. 131–45). Champaign, IL: Palgrave Macmillan.

Huws, U. (2014). *Labor in the global digital economy: The cybertariat comes of age.* New York, NY: NYU Press.

Lent, R. (2018). Future of work in the digital world: Preparing for instability and opportunity. *The Career Development Quarterly*, **66**(3), 205–19.

MacCrory, F., Westerman, G., Alhammadi, Y. and Brynjolfsson, E. (2014). Racing with and against the machine: Changes in occupational skill composition in an era of rapid technological advance. In *Proceedings, Thirty Fifth International Conference on Information Systems, New Zealand.* Retrieved from https://pdfs.semanticscholar .org/164e/93f0d99852a2b8474c9c0c902eb00807a379.pdf.

Manyika, J., Lund, S., Chui, M., Bughin, J., Woetzel, J., Batra, P., Ko, R. et al. (2017). Jobs lost, jobs gained: Workforce transitions in a time of automation. San Francisco, CA: McKinsey Global Institute.

Miller, R. (2018). *Transforming the future: Anticipation in the 21st century.* Paris: UNESCO Publishing.

Mishel, L. and Bivens, J. (2017). The zombie robot argument lurches on: There is no evidence that automation leads to joblessness or inequality. Economic Policy Institute. Retrieved from Economic Policy Institute website: https://www.epi .org/publication/the-zombie-robot-argument-lurches-on-there-is-no-evidence-that -automation-leads-to-joblessness-or-inequality/.

Neufeind, M., O'Reilly, J. and Ranft, F. (2018). *Work in the digital age: Challenges of the fourth industrial revolution.* London: Rowan & Littlefield.

Schwab, K. (2016). *The fourth industrial revolution.* Geneva: World Economic Forum.

Spencer, D.A. (2018). Fear and hope in an age of mass automation: Debating the future of work. *New Technology Work and Employment*, **33**(1), 1–12.

Spreitzer, G.M., Cameron, L. and Garrett, L. (2017). Alternative work arrangements: Two images of the new world of work. *Annual Review of Organizational Psychology and Organizational Behavior*, **4**(1), 473–99.

Standing, G. (2011). *The precariat: The new dangerous class.* London: Bloomsbury Academic.

Susskind, R. and Susskind, D. (2016). *The future of the professions: How technology will transform the work of human experts.* Oxford: Oxford University Press.

Tytler, R., Bridgstock, R.S., White, P., Mather, D., McCandless, T. and Grant-Iramu, M. (2019). *100 jobs of the future.* Ford Australia. Retrieved from https:// 100jobsofthefuture.com/report/.

World Economic Forum (2016). The future of jobs: Employment, skills and workforce strategy for the fourth industrial revolution. Global Challenge Insight Report. Retrieved from http://www3.weforum.org/docs/WEF_Future_of_Jobs.pdf.

World Economic Forum (2018). The future of jobs report 2018. Retrieved from http:// reports.weforum.org/future-of-jobs-2018/.

Index